LOCKHEED MODEL L-200 *CONVOY FIGHTER*
THE ORIGINAL PROPOSAL AND EARLY DEVELOPMENT OF THE XFV-1 SALMON

PART 1

Jared A. Zichek

RETROMECHANIX PRODUCTIONS

First published in the United States of America in 2017 by Jared A. Zichek, 12615 North Wildwood Point Road, Hauser, Idaho 83854, USA

E-mail: editor@retromechanix.com

©2017 Jared A. Zichek

All rights reserved. All featured text and images are copyright 2017 their respective copyright holders. No part of this publication may be reproduced, stored in a retrieval system, or transmitted in any form by any means electronic, mechanical or otherwise without the written permission of the publisher.

ISBN: 978-0-9968754-4-8

www.retromechanix.com

All images in this publication are scanned from documents held by National Archives II, College Park, MD unless otherwise indicated. All color profile artwork is ©2017 Jared A. Zichek. Special thanks to John Aldaz for the photos of his L-200-1 display model. Printed in USA.

Front Cover: At the top is a contemporary artist's impression of the Lockheed Model L-200 Navy VF (Visual Fighter) Convoy Fighter proposal of 1950, which ultimately led to the XFV-1 Salmon. The proposal differed from the XFV-1 in several respects, most notably in having a tail with only three surfaces and an eight-bladed propeller. Below this is a speculative color profile of the L-200-1, the tactical version of the Convoy Fighter. The overall Glossy Sea Blue scheme was the standard for most US Navy aircraft of the early 1950s; markings are inspired by the actual XFV-1 prototype. (Note: this color profile is not to the same scale as those on the back cover).

Introduction

The Lockheed Model L-200 was one of five proposals submitted to the US Navy Convoy Fighter competition of 1950, which called for a high performance turboprop VTOL fighter to protect convoy vessels from attack by enemy aircraft. Other participants included Convair, Goodyear, Martin and Northrop, whose proposals have been covered in previous volumes. This book reproduces part of Lockheed's original proposal for what became the XFV-1 Salmon; the second volume will cover the remainder of the proposal along with the aircraft's early development with NACA.

For those unfamiliar with the history of the concept, the idea of a turboprop tailsitter fighter emerged in the late 1940s, with the US Navy Bureau of Aeronautics (BuAer) beginning to seriously examine the feasibility of developing a vertical takeoff and landing (VTOL) tailsitter aircraft to protect convoys, task forces, and other vessels. These specialized interceptors would be placed on the decks of ships to provide a rapid defensive and reconnaissance capability until conventional carrier-based fighters could arrive and assist. The Battle of the Atlantic was fresh in the minds of Navy planners, who were concerned that the Soviets would engage in a similar campaign against merchant shipping if the nascent Cold War erupted into open conflict. BuAer's interest in a VTOL tailsitter fighter coincided with the development of new turboprop engines which provided enough horsepower to make the concept a reality.

BuAer's *Outline Specification for Class VF Airplane (Convoy Fighter) OS-122* was dated 10 July 1950. It listed the requirements for such an aircraft along with a scale demonstrator to verify the soundness of the concept. The document was distributed to the major aircraft manufacturers of the day, with the aforementioned companies responding in late November 1950. The products of this competition, the Convair XFY-1 Pogo and Lockheed XFV-1 Salmon, never made it beyond the prototype stage, as they proved to be very difficult to land, suffered from power plant reli-

▲ 1

1) **Cover to the Lockheed Convoy Fighter preliminary design summary report dated 10 November 1950.**

ability issues, and were eclipsed in performance by contemporary jet fighters. They became historical curiosities, regularly making the list of world's worst/strangest aircraft, an ignominious end for a program initiated with such high hopes at the beginning of the 1950s.

L-200 Navy VF Convoy Fighter Design Features

In addition to general design work to fulfill the requirements of BuAer Specification OS-122, Lockheed instigated and completed a research program which included powered wind tunnel tests, a free flight powered model, and a movable cockpit mock-up. These programs assisted in evaluation the magnitude of the problems in making a vertical rising high performance airplane and suggested several possible solutions for the major design problems. Since the problem was unconventional and would involve a considerable research program before a successful airplane was developed, it was difficult to select any one complete design which fulfilled the

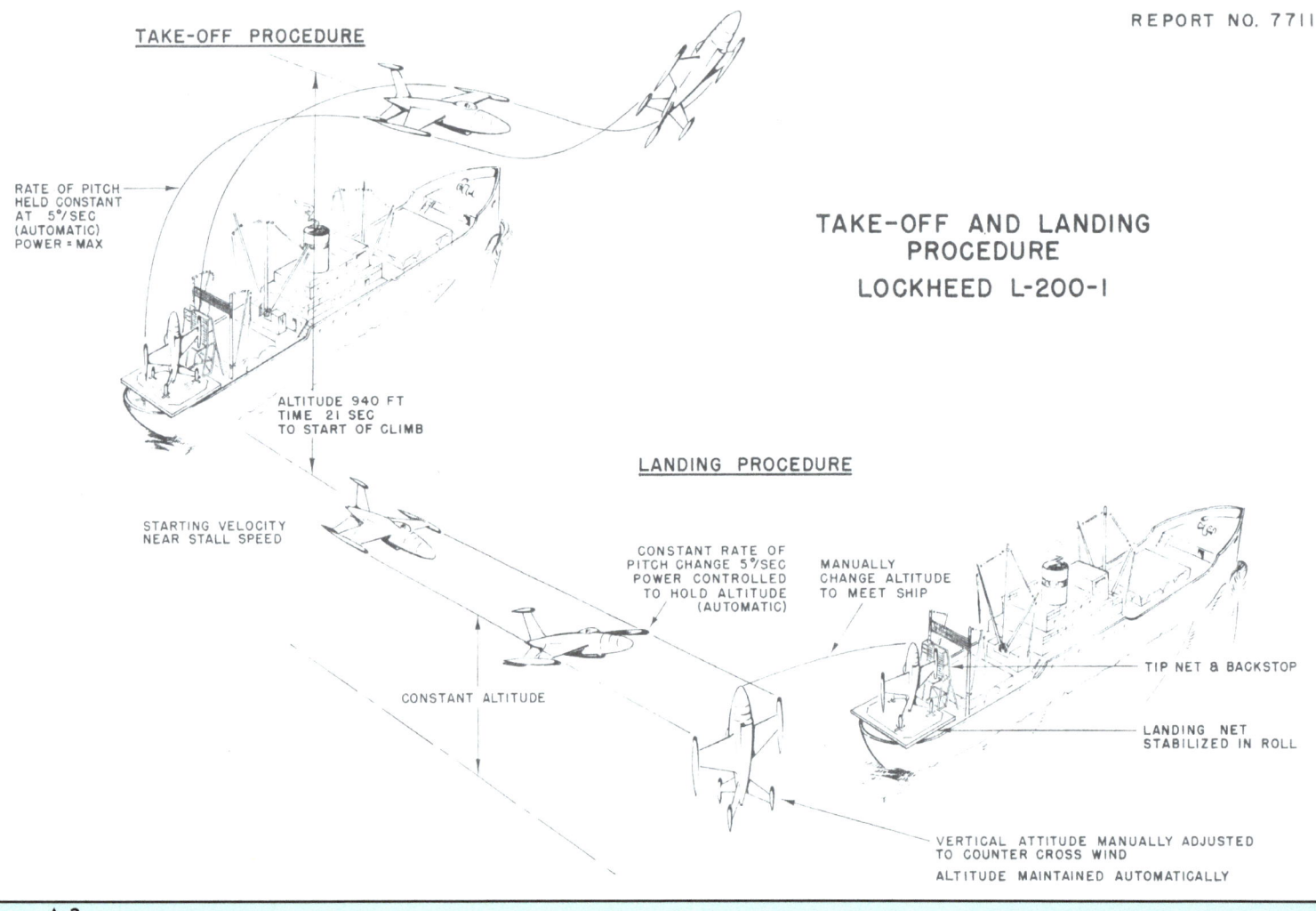

▲ 2

specification ideally in all categories. As a result, this proposal, although showing one final design, presented alternate designs which had unique advantages. These, for the most part, were accompanied by unknown features which required further research to confirm their feasibility. Lockheed hoped to study these alternate design ideas and institute research programs during the first phases of a further design investigation in order to achieve additional desirable features in the final airplane.

2) Illustration of the final takeoff and landing procedures which evolved from the study of the many problems encountered in this unconventional procedure. Although the ideal solution for some of these problems may not have been achieved in this system, Lockheed believed that the general solution almost completely fulfilled all the requirements and accounted for more contingencies than any other alternate design studied by the company.

3) Graphic of the performance of the L-200 on interception and search missions using the deck of a tanker or cargo vessel as a landing and takeoff area.

4) Table comparing the resulting performance of the Lockheed L-200-1 with BuAer requirements; according to the company, the airplane closely approached or exceeded all of the performance items required in the initial specification. The interception of a 450 knot bomber was possible at 35,000 ft of altitude with an early warning distance as low as 146 nautical miles from the convoy. This warning distance was entirely possible using the L-200-1 airplane for early warning convoy patrol.

Operational Features of the L-200

Flight Performance

The high speed performance of the airplane was equivalent to the specific requirements in that a Mach number of 0.94 was achieved at an altitude of 35,000 ft. This performance, of course, did not indicate the complete capabilities of the airplane at other altitudes nor did it show the maneuverability or cruise performance of the airplane under other circumstances than those outlined in OS-122. Although the estimated performance equaled or exceeded existing carrier-based fighters, the primary and unique performance advantage of the airplane was its

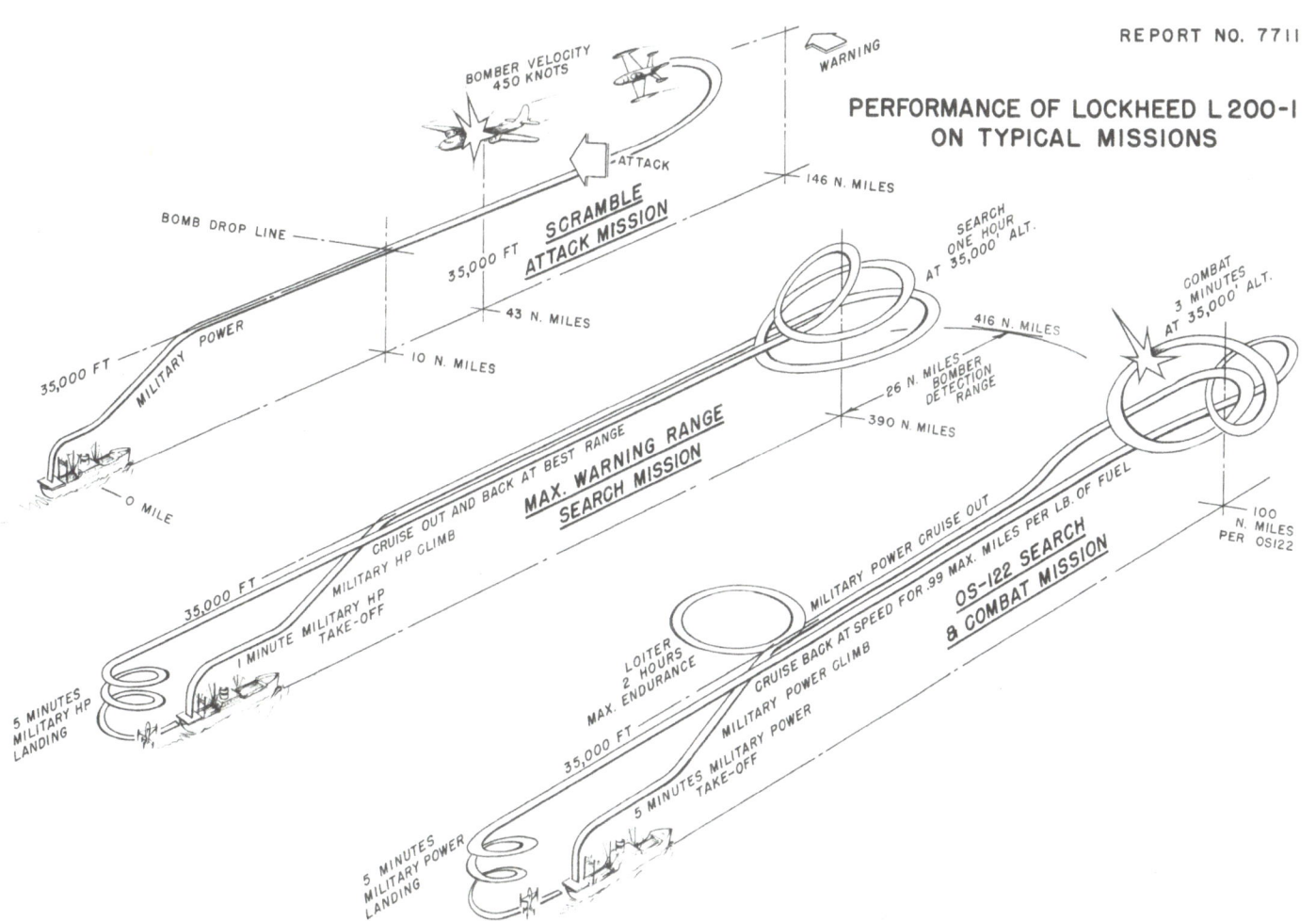

▲ 3

ability to land and take off in a vertical direction.

Pilot Operating Procedures

Takeoff and Landing Procedures

In the study of the vertical rising portions of the airplane operation, a mock-up was made of the proposed cockpit, a complete study was made of Navy information which accompanied Specification OS-122, and a trip was made to the Naval Aircraft Development Center in Johnsville to inspect the mock-up which had resulted from Navy studies of this problem. Although many alternate positions of the pilot were investigated, it was finally concluded that the pilot's position shown on p. 5 was the most desirable. In this position, the pilot's seat was rotated about an axis just below the pilot's elbow so that the pilot's head was raised toward the instrument panel when the airplane reached its vertical attitude. In this position, no change in the controls of the airplane nor rearrangement of instruments appeared to be required and, as a result, an almost normal cockpit was achieved.

It was felt, however, that backward vision from this pilot's position was extremely

COMPARISON OF REQUESTED PERFORMANCE AND ESTIMATED L-200-1 PERFORMANCE		
	Spec. OS-122	L-200-1 Estimated
Maximum Gross Weight	16,000	15,600
Engine	T40A-8	T40A-8
Armament	4-20mm cannon	4-20mm cannon
Ammunition	150 rds/gun	150 rds/gun
Radar	APQ/42	APQ/42
(Alternate Radar)	-	Westinghouse Advanced Beam Attack Fire Control System
Fuel Capacity	-	508 gal.
Combat Weight (-40% Fuel)	-	14,380 lbs.
Loiter Duration per OS-122	2 hours	2 hours
Combat Ceiling (500 ft./min.climb)	45,000 ft.	48,500 ft.
Acceleration in Transition	5 ft./sec.^{2}	7.3 ft./sec.^{2}
Time to Climb to 35,000 ft.	4.5 min.	4.9 min.
$V_{max.}$ at sea level, Military Horsepower	-	526 kn
$V_{max.}$ at 20,000 ft., Military Horsepower	-	563 kn
$V_{max.}$ at 35,000 ft.(Combat Weight)	540 kn	548 kn
Horizontal to Vertical Transition Time	-	15 sec. approx.
Take-Off Transition Time	-	21 sec.
Altitude Change in Transition to Vertical	500 ft. max.	None

▲ 4

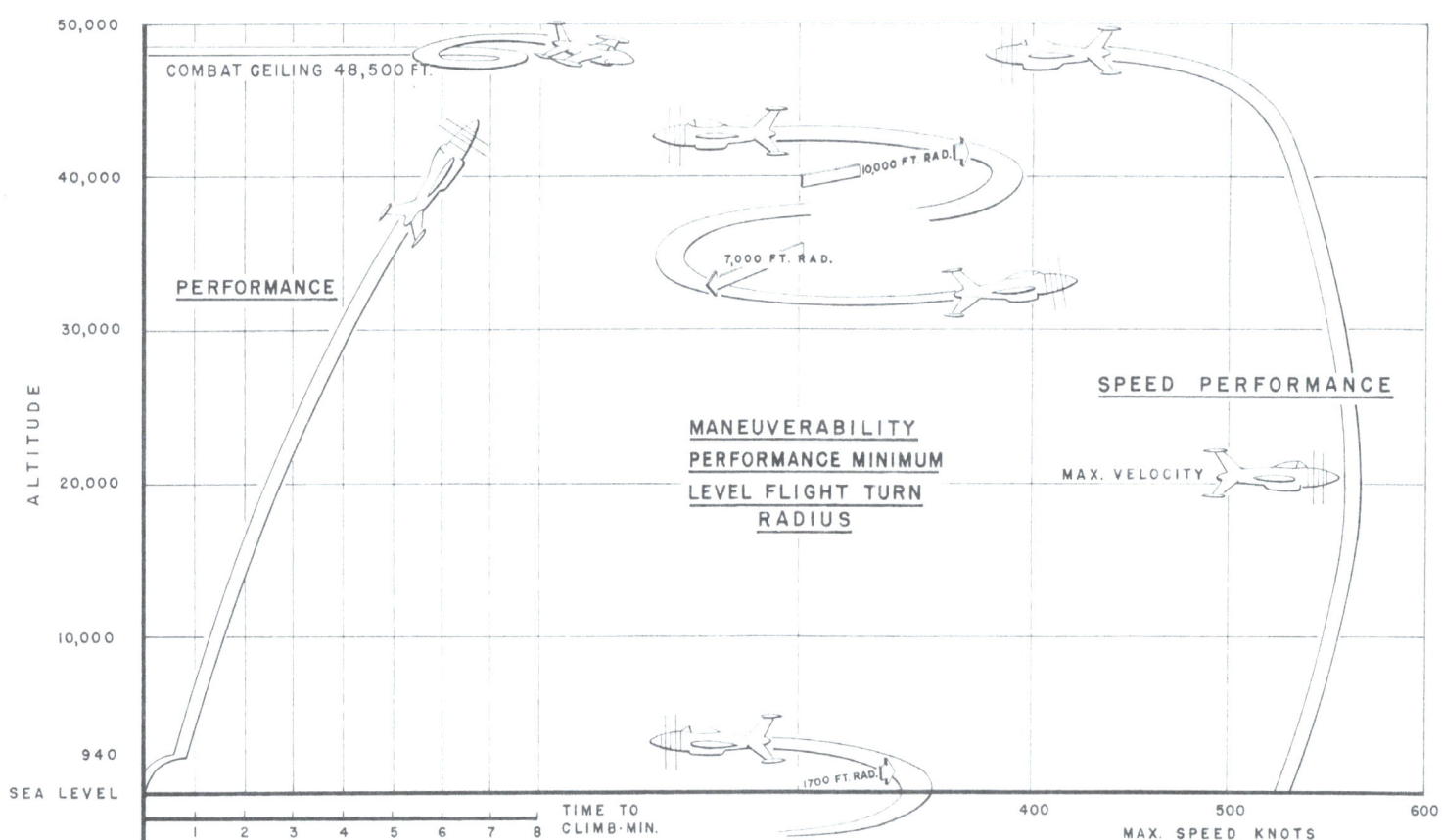

5) This summary of combat performance was created to illustrate the flexibility and utility of the L-200 Convoy Fighter. On paper, it was competitive with conventional carrier-based types while offering the unique capability of taking off and landing vertically.

uncomfortable and that somehow landing assistance had to be given to the pilot during his final touchdown maneuver so that backward vision was normally unnecessary. This led to the inclusion in the landing scheme of a landing signal officer (LSO) stationed on the landing platform on board the cargo vessel. Properly located, the LSO could adjust the vertical height of the airplane by signals to the pilot, accounting for the motion of the landing platform due to the sea condition. Furthermore, if properly located, it only became necessary for the pilot of the airplane to look directly sideways out of the cockpit, a direction which afforded maximum visibility, a maximum comfort, and, due to the location of the wing tip on the airplane, a visual reference as to the attitude and motion of the airplane itself.

With these major conclusions in mind, the deck facilities and procedures for landing and takeoff finally evolved. The major landing area consisted of a tautly drawn cable net with approximately a six inch mesh. A net was used so that many landing holes would be available for the tip spikes and no obstruction to the airflow past the tail would result from the ground plane effect on the propeller slipstream. This net was mounted on a platform and raised above the deck of the cargo vessel by means of support arms with bearings at each end. These arms were so arranged that the entire platform could be moved in a roll direction and stabilized to suit any roll of the vessel, achieving a consistently level platform no matter what the sea state. Since the pitch angle of the ship was relatively small and could be accounted for in angular tolerance between the airplane and net, no effort was made to achieve a level attitude in pitch. Lockheed noted, however, that the support arms for the landing platform were so arranged that the motion, which maintained the platform level, moved the platform in a direction to counteract the sideways translation of the deck due to the ship's roll. Thus, the platform itself remained

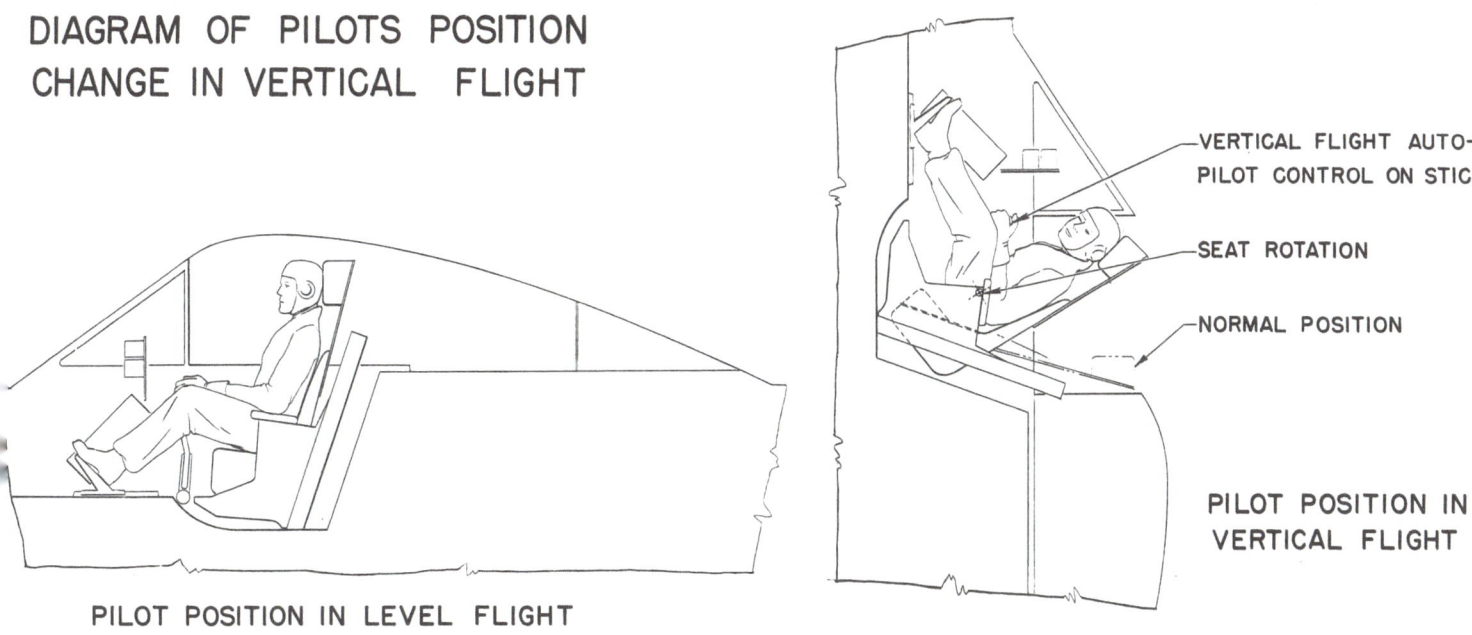

DIAGRAM OF PILOTS POSITION CHANGE IN VERTICAL FLIGHT

PILOT POSITION IN LEVEL FLIGHT

PILOT POSITION IN VERTICAL FLIGHT
- VERTICAL FLIGHT AUTO-PILOT CONTROL ON STICK
- SEAT ROTATION
- NORMAL POSITION

▲ 6

level, rising up and down only approximately 18 inches due to roll and translated from side to side only ±44 inches under the most severe roll conditions anticipated by the Navy. This net itself under ideal conditions could be the entire landing facility for the airplane if accurate flight control could be assured and if it could be assumed that the LSO and the pilot together could accurately position the airplane in spite of a pitching and rolling deck without having the airplane drop into the net with any major translational relative velocity. With further development of the landing scheme this could have been eventually achieved but it was felt that such an assumption was extremely dangerous at the time. As a result, additional features were added to the landing platform which provided further stability for the airplane once the landing had been effected, and permitted large tolerances in the translational velocity at the time of touchdown. These facilities consisted of a tip net of similar construction to the main landing platform and a "tip backstop." With these additions attached to the stabilized platform, the landing facilities were complete. The procedure for the pilot during landing was as follows:

6) Lockheed studied several options for the pilot position before settling on this design, where the seat rotated about an axis just below the pilot's elbow so that his head was raised toward the instrument panel when the airplane reached the vertical attitude.

Landing Procedure

1. The airplane approached perpendicular to the cargo vessel on the port side in a level flight attitude making a transition to vertical flight to the stern and above the landing vessel.
2. The pilot set the translational velocity required to hold against the cross wind into the autopilot control through the special vertical flight autopilot controller on the end of the control stick.
3. Flying parallel to the ship, the pilot closed distance in both altitude and position by decreasing altitude and translating the airplane sidewards toward the landing platform. This was done by reducing power and moving the vertical flight autopilot controller to the left.
4. With the airplane in close position in altitude, being slightly to stern of the vessel, flight direction responsibility was taken over by the LSO on the landing platform.
5. Under LSO direction, the airplane was then translated sideways directly over the landing platform at sufficient altitude to clear the highest plunge position of the ship stern. In this position, the pilot was ready for cut when the landing platform was at its

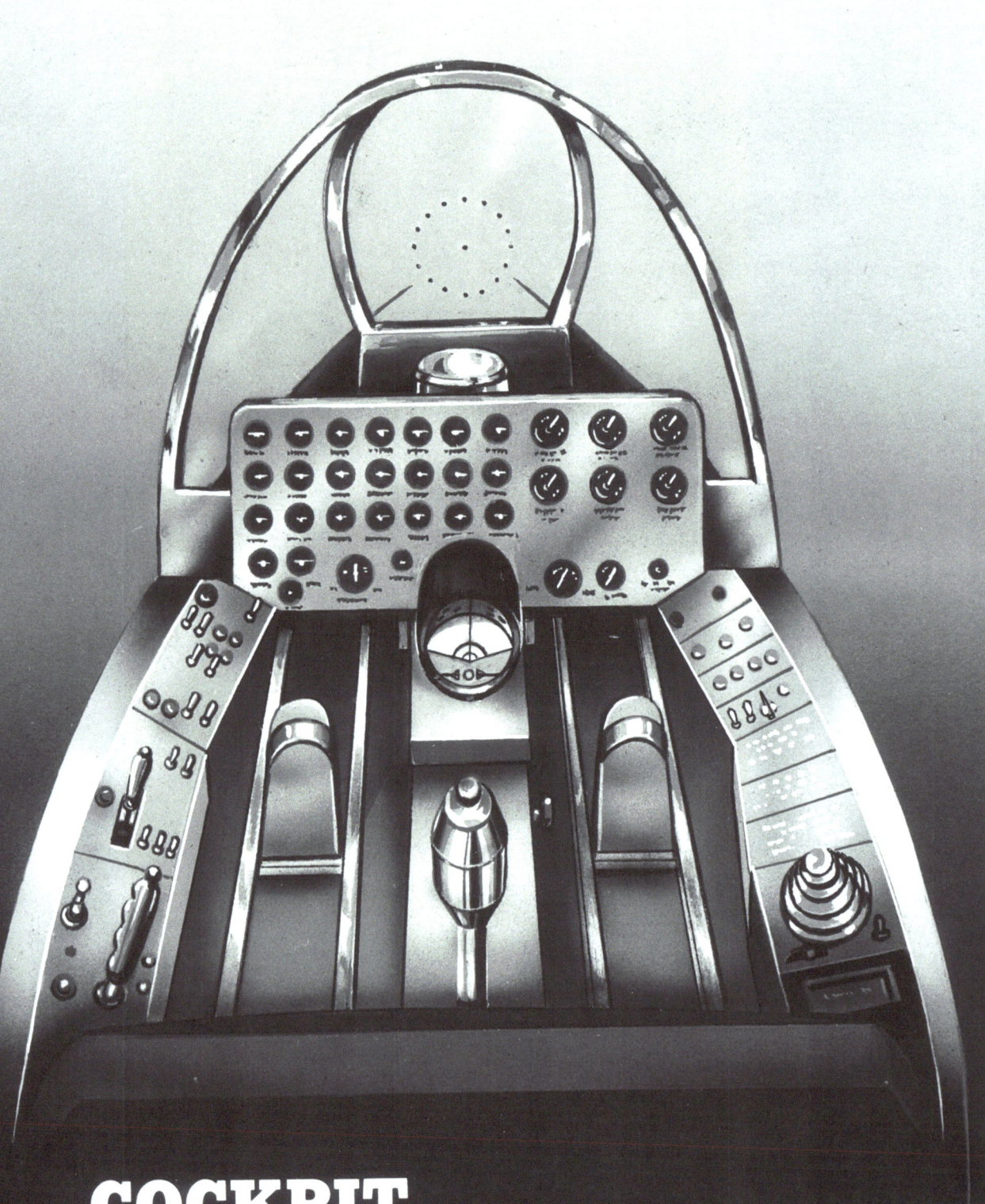

COCKPIT
FIRE CONTROL, RADAR INDICATOR AND MK6 MOD. 1 OPTICAL SIGHT

EMERGENCY PROVISIONS

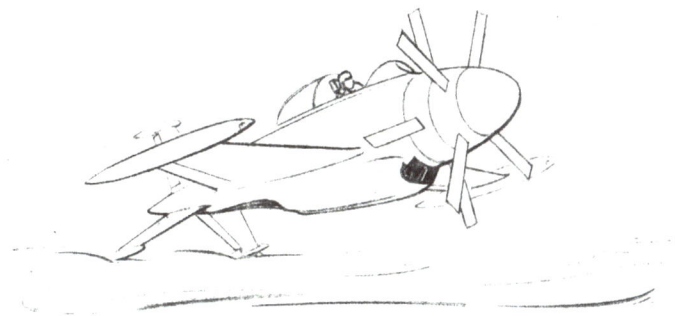

(A) JETTISON SEAT EJECTION

(B) IN FLIGHT OPENING CANOPY

(C) DITCHING DOORS

(D) BRAKING FOR PROPELLERS

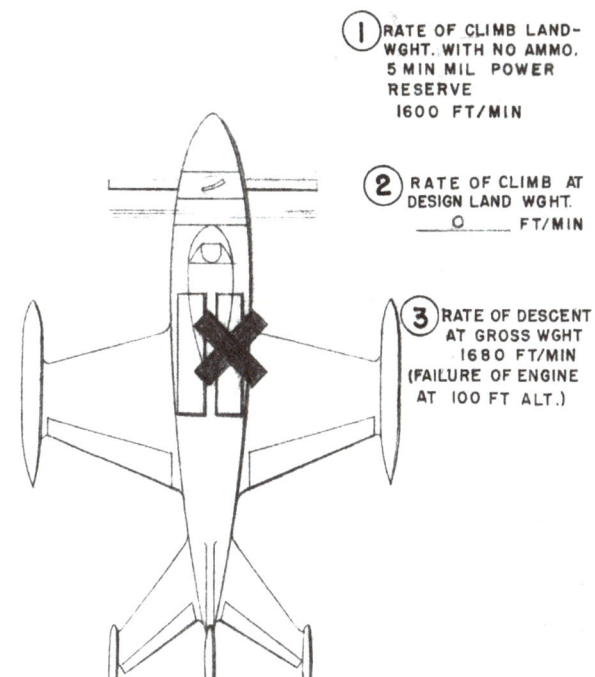

(E) ONE ENGINE OUT HOVERING PERFORMANCE

① RATE OF CLIMB LAND-WGHT. WITH NO AMMO. 5 MIN MIL POWER RESERVE 1600 FT/MIN

② RATE OF CLIMB AT DESIGN LAND WGHT. 0 FT/MIN

③ RATE OF DESCENT AT GROSS WGHT 1680 FT/MIN (FAILURE OF ENGINE AT 100 FT ALT.)

▲ 8

proper position (slightly below crest and rising).

6. Just prior to cut, the pilot was given the signal to close in a sideways direction against the tip backstop which relieved him of the responsibility for accurate sideward positioning exactly at cut. The tip pod of the airplane struck the shock absorbing backstop which was equipped with rollers and, at the cut signal, the airplane dropped into the landing net standing on the three shock absorbing tail stands and pierced the tip net with the aft point of the tip nacelle. Thus anchored, the airplane could not tip even under the most severe conditions of ship pitch or roll, even though the platform stabilization devices were turned off.

7. In the event of difficulty, the pilot could elect a wave-off. Since the tip stop was located on the forward side of the platform where the normal ship's obstructions such as masts, booms, etc. were located, this structure did not constitute an additional hazard. Wave-off could be taken either up or to either side or aft.

Takeoff Procedure

1. With the engines running and the airplane mounted in a vertical position on the landing and takeoff platform, engine checkout was performed, running to maximum thrust, with the airplane anchored through a tail cable to a catapult holdback.
2. After check-out by the pilot and set-

7) The cockpit of the L-200 had all flight instrumentation grouped in its standard arrangements, with all engine instrumentation carefully grouped on the left of the panel, and the APQ/42 radar screen located directly in the center and immediately below the gunsight.

8) In the event of an emergency, the Convoy Fighter's ejection seat could be actuated in either the vertical or horizontal position of the airplane,

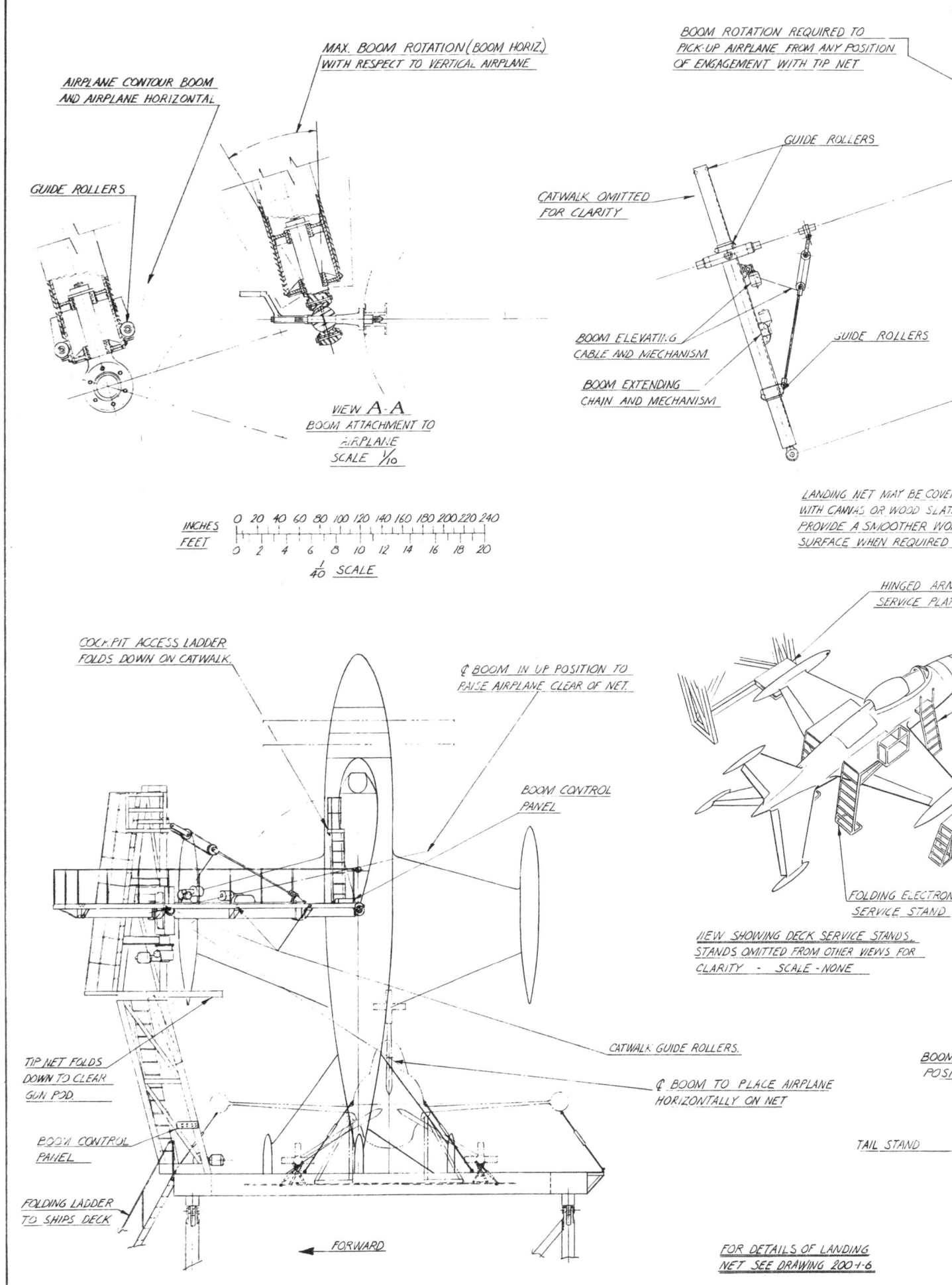

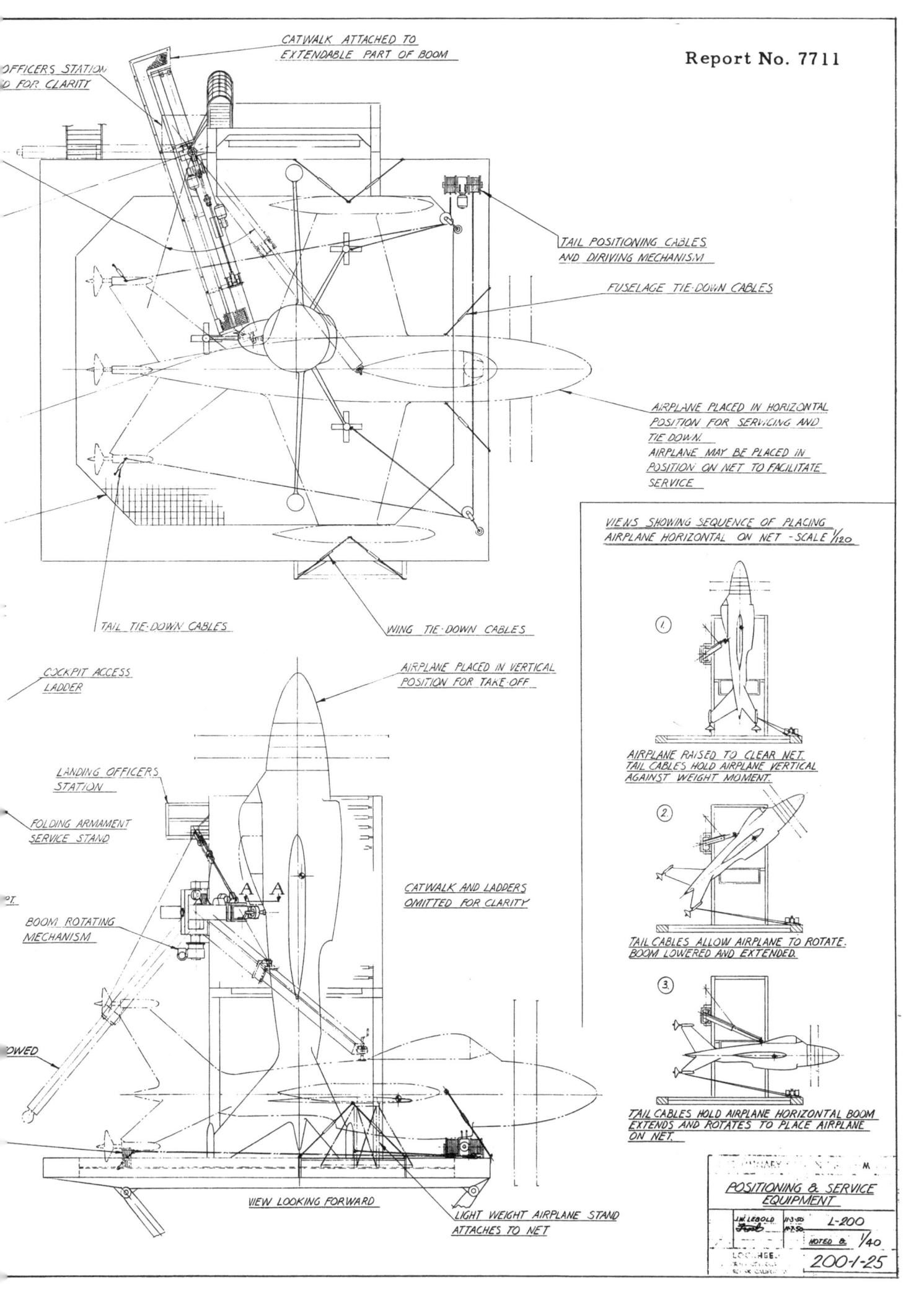

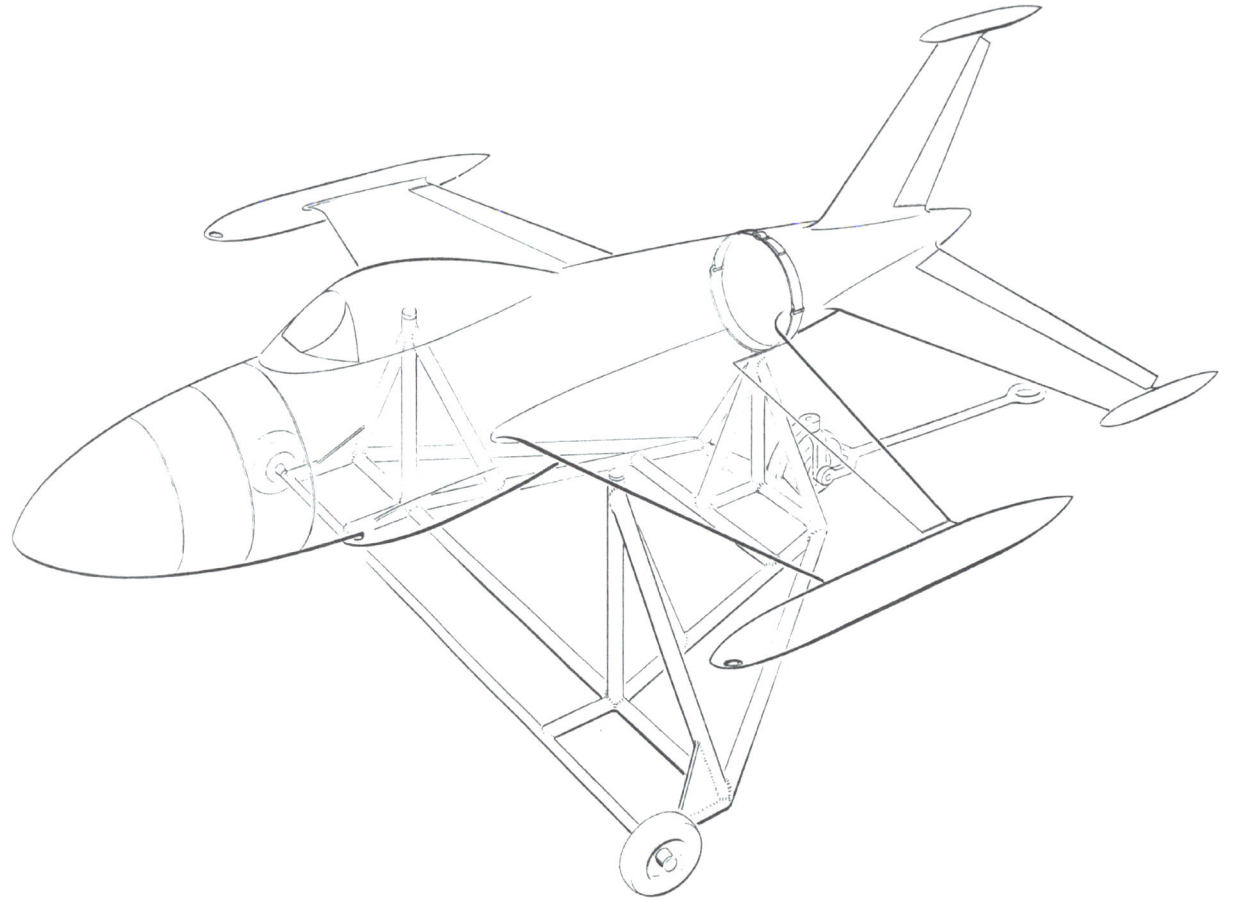

AIRPLANE ON SERVICE CART

▲ 10

3. ting in of roll, pitch and yaw stabilization of the autopilot about a vertical axis, the pilot signaled the LSO that all was in readiness for takeoff.
3. The LSO armed the breakaway feature of the hold-down cable (similar, but designed for more accurate loads, to break the ring of a catapult holdback) and the pilot opened the throttles to the maximum thrust condition. (A plain break ring could have been satisfactory if full power run-up was not required). When takeoff thrust was achieved, the break ring permitted the airplane to leave the deck (the tip net, which was retractable, was pulled away as soon as the tail anchor cable was secured).

These procedures for takeoff and landing may not have represented the ultimate in simplicity but it was felt that they satisfactorily accounted for the foreseeable contingencies and permitted sufficient inaccuracy in flying the airplane to prove the feasibility of the entire vertical rising scheme. Alternate landing methods were studied, each of which had unique features in permitting additional tolerances in flying the airplane at the expense of deck complications and which permitted considerable deck simplification if such tolerance was not required. Lockheed pointed out that one of the disadvantages of the plain net for landing was that the stability of a relatively normal airplane landing on its tail was small in a tip-over direction, permitting only very low translational velocities at touchdown. This was improved by the addition of a backstop and tip net in this proposal but could have been

Previous Spread

9) Detailed drawings of the major components of service equipment required for deck handling and servicing of the L-200.

This Page

10) Illustration of the cart which served as a maintenance or moving stand when the airplane was away from its landing platform.

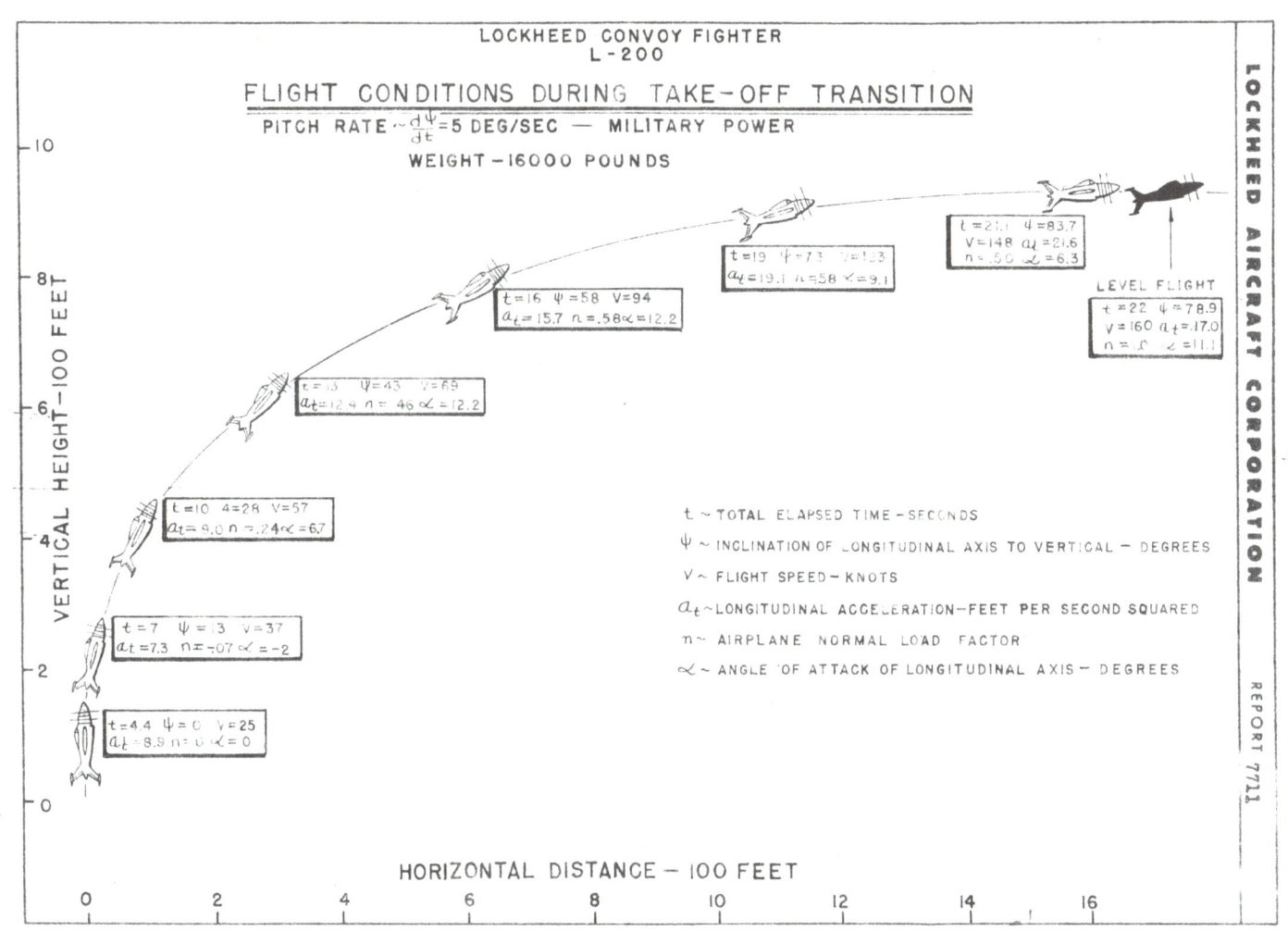

solved by the following schemes:

1. Extreme wide span or additional landing prongs on the airplane tail. This resulted in a large weight penalty which was not permitted on the airplane.
2. Wheels on the tip of the tail, which were fully swiveling so that no side reaction could be obtained on landing, required that side translation be stopped by a tail hook or dangling hooks from the center of gravity (CG) of the airplane. This scheme eliminated the use of a net which made it necessary to have the tail close to the platform and it was doubtful whether proper control could was achievable if a ground plane was this close to the tail controls. Furthermore, a large weight penalty resulted on the airplane.
3. A wide spread of the landing prongs could be achieved by designing a tail-first or all-wing airplane where the landing was effected on the wing tips rather than on the tail. This scheme had many possibilities but the airplane designs resulting from such a procedure had many unknown aerodynamic qualities that could not be investigated soon enough for a reasonable proposal to be made. Some of these alternate designs are shown on pp. 16-17.

Transition Phases of Flight

Transition from horizontal to vertical flight was relatively simple operation if the proper autopilot was installed. With this autopilot, the procedures for takeoff transition and landing transition were evolved. For takeoff it was only necessary to set a given rate of pitch into the autopilot and open the throttles to their maximum

11) In its study of takeoff transition, Lockheed recommended a 5° per second pitch rate be used which resulted in transition in 21 seconds using 940 ft of altitude.

LOCKHEED L-200

TRANSITION TO HOVERING ATTITUDE

RATE OF PITCH, $d\alpha/dt$ = 3 DEG./SEC.

Time, Sec.	0	5	10	15	20	25	31
Rate of Pitch, Deg/Sec.	3	3	3	3	3	3	0
Angle of Attack, Deg.	14	29	44	59	73	88	90
Velocity, Knots	147	123	90	59	33	21	0
Percent Power	0	18	33	53	64	66	68
Control Setting, Deg.	-8	-13	-15	-10	-2	+8	0
Distance Covered, Ft.	0	1190	2180	2830	3200	3400	3500
Deceleration, g's	-.18	-.33	-.32	-.36	-.18	-.19	0

RATE OF PITCH, $d\alpha/dt$ = 5 DEG/SEC.

Time, Sec.	0	5	10	15	24
Rate of Pitch, Deg/Sec.	5	5	5	5	0
Angle of Attack, Deg.	14	39	64	89	90
Velocity, Knots	147	114	67	20	0
Percent Power	0	10	45	66	68
Control Setting, Deg.	-8	-18	-9	+9	0
Distance Covered, Ft.	0	1140	1970	2400	2600
Deceleration, g's	-.18	-.48	-.51	-.30	0

▲ 12

takeoff power. This rate of pitch was determined by calculation to be such that the transition was made to best climb speed in the minimum time without exceeding any aerodynamic limitations on the airplane. The rate of pitch chosen for the presentation was 5° per second which resulted in a level flight speed slightly above power-off stall speed when level flight had been reached. Optimum methods for interception could be calculated when more was known of the airplane characteristics. Adjustment of this rate of pitch change could be made simply by overcontrolling the autopilot with manual control or by readjusting the initial setting with the standard autopilot controller. The stick autopilot controller was not used during this portion of the flight transition but the hand could be held on the stick since the rate of pitch normally did not need to be readjusted and was automatically cut out when the airplane achieved a horizontal attitude.

The landing transition was simple since it could be achieved by setting a constant rate of pitch in the autopilot. This rate incidentally was approximately the same in a pitch-up as was the takeoff rate in a pitch-down direction. In this case, the maximum pitch rate was dictated by a desire not to have the airplane traveling too fast in a horizontal direction when the vertical attitude was reached. This seriously impaired control. In order to limit the altitude of the airplane during pitch-up, it was possible to set into the power controls a zero rate of change of altitude from an aneroid rate of change signal so that during the maneuver, the power was continuously adjusted to hold a constant altitude. Calculations showed that a transition to vertical flight could be made with no altitude change. This made possible an approach to the cargo vessel at relatively low altitudes under very poor visibility. Once a vertical position was reached, an entirely different autopilot controller was used since the airplane became stabilized in a vertical attitude by the vertical gyro which was included in the

12) Lockheed considered a 5° per second pitch rate transition optimal in that it required 15 seconds and resulted in a translational velocity when the airplane attitude had reached the vertical of 31.4 kts.

LOCKHEED L-200
SWEPT WING

STEADY STATE TRANSLATION

WEIGHT = 16,000 LB.

Angle of Attack	90°	80°	70°	60°	50°
Without Flaps:					
Velocity, Knots	0	14	28	42	56
Stabilizer Angle*	4°	2.7°	0.5°	-1.7°	-4.0°
Percent Power	68%	67%	65%	59%	50%
With Flaps:					
Velocity, Knots	16	27	36	46	58

*Wind Tunnel data in terms of stabilizer angle for convenience

Angle of Yaw	0°	10°	20°	30°	40°
Velocity, Knots	0	43	86	-	-

LOCKHEED AIRCRAFT CORPORATION REPORT NO. 7711

autopilot system. At this point, roll stabilization was left in and rate stabilization was also utilized for both yaw and pitch, but the separate vertical flight controller on the end of the stick was utilized as described previously in the landing procedure.

Cruise and Search Procedures

The cockpit of the Lockheed L-200-1 was arranged to provide the best possible facilities for comfortable prolonged operations for the use of the APQ/42 radar and fire control system. (Similar facilities also applied to the advanced Westinghouse radar and fire control system proposed as an alternate by the company). All flight instrumentation was grouped in its standard arrangements, all engine instrumentation was carefully grouped on the left of the panel, and the radar screen was located directly in the center and immediately below the gunsight. In this position, the radar scope was easily seen in a comfortable seating position by the pilot, either for search or combat. Directly to the right of the right arm rest was the standard autopilot controller which was in a position within easy reach of the pilot's right hand and approximately in the opposite position in the cockpit from the power handle on the left. At cruise or search altitudes, the radar swept the sky within ±75° from the centerline of the airplane. The search range was expected to be 26 nautical miles on a B-29 type of target. Lockheed determined that the L-200-1 could act as an aerial search airplane for convoy purposes and that a search radius of 406 nautical miles could be achieved with one hour loiter on patrol and with a 10% reserve after returning to the convoy. This radius was actually not usable with the listed communicating equipment since maximum communicating range was approximately 120 nautical miles. This calculated early warning radius could have been used, however, if different communicating equipment were installed. This facilitated convoy protection

13) This illustration demonstrates the airplane attitude required to translate in the fore, aft and side directions as derived from wind tunnel test data.

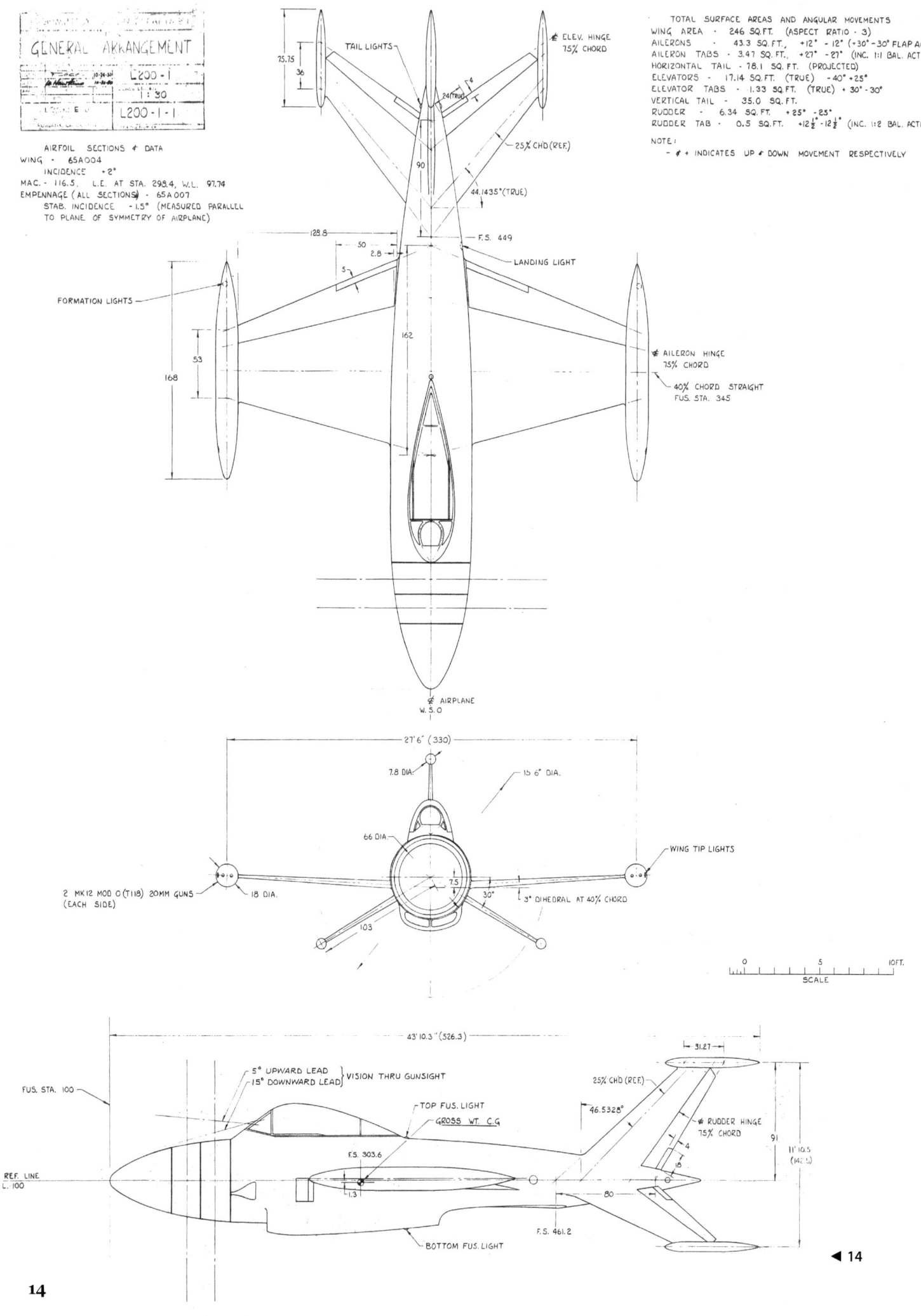

since two airplanes (or groups) could be airborne simultaneously having taken off at different times, one searching, the other being available to join combat in the event that the early warning airplane discovered a target. If no target was discovered within the search time of the first group, the second one could take over the search mission and another one could be made airborne to stand by for combat.

Combat Procedure

As previously noted in the Flight Performance Section, the maneuverability of the airplane was excellent at the altitudes where interception could be expected. The combat equipment comprised of the APQ/42 radar combined with Mark VI Mod. I gunsight and the four 20 mm cannon made the airplane an excellent interceptor for use against targets of the B-29 type. For using the radar for gunlaying, the radar controls were located within easy reach of the pilot on the left console.

If a target was located, the convoy was notified and an approach to a pursuit course interception was initiated. The run-in was made, keeping the target boresighted until lock-on range of the target was achieved, as noted on the radar scope range lines. At this point, the radar was locked on the target by means of the control handle on the left side of the cockpit at which point the radar stayed locked on the target until such time as the target went out of range or firing position was reached.

It was the conclusion of Lockheed's studies that the four 20 mm cannon probably were not the ultimate armament of the airplane nor was the APQ/42 the ultimate gunlaying radar. Therefore, a study was made which showed an alternate radar installation which combined with alternate tip tanks containing 24 folding fin rockets each. With this armament arrangement, it was possible to achieve a beam attack rather than a pursuit attack course, thereby protecting the pilot of the fighter and improving the chance of kill on the bomber. These provisions were easily incorporated in the airplane and it was believed than any design which did not have the possibility of carrying rocket armament unnecessarily restricted the fire power of an airplane designed to be operational in 1954 or later.

Emergency Procedures

One of the problems which became extremely acute in a vertical rising airplane was the manner in which the pilot could be safeguarded in the event of inadvertent power failure. An ejection seat was incorporated which could be actuated in either the vertical or horizontal position of the airplane, ditching doors were provided to ensure the airplane's stability in the ditching operation and to prevent water from entering the cockpit or destroying the airplane by ramming the inlet ducts of the engine. An in-flight movable canopy was provided in order to ensure the pilot of an exit in the event that the airplane dropped in the water and, finally, the airplane was kept at a light enough weight that hovering performance was feasible at the normal landing weight with one power section dead. Since single engine performance was possible in hovering, special care was expended in the systems which were common to both engines to ensure that no failure on one power section would be transmitted to the other power section. The two inlet ducts were separated, the oil cooling systems were separate with separate tanks, and the duct closure doors could be operated individually to close one power section and prevent destructive windmilling.

Facilities for Maintenance and Service

Lockheed studied many alternate schemes in achieving its final configuration of service and maintenance facilities, most of which consisted of utilizing the hoisting facilities already available on cargo vessels. The major disadvantage of this type of arrangement was that cables had to be attached to the airplane at a long distance from the deck which required stands for reaching the hoist fittings on the airplane, almost as complicated as the mechanism which permitted

14) Three-view of the Lockheed L-200-1 Convoy Fighter proposal; the key difference between this configuration and the XFV-1 Salmon concerned the tail, with the former only having three surfaces and the latter having four.

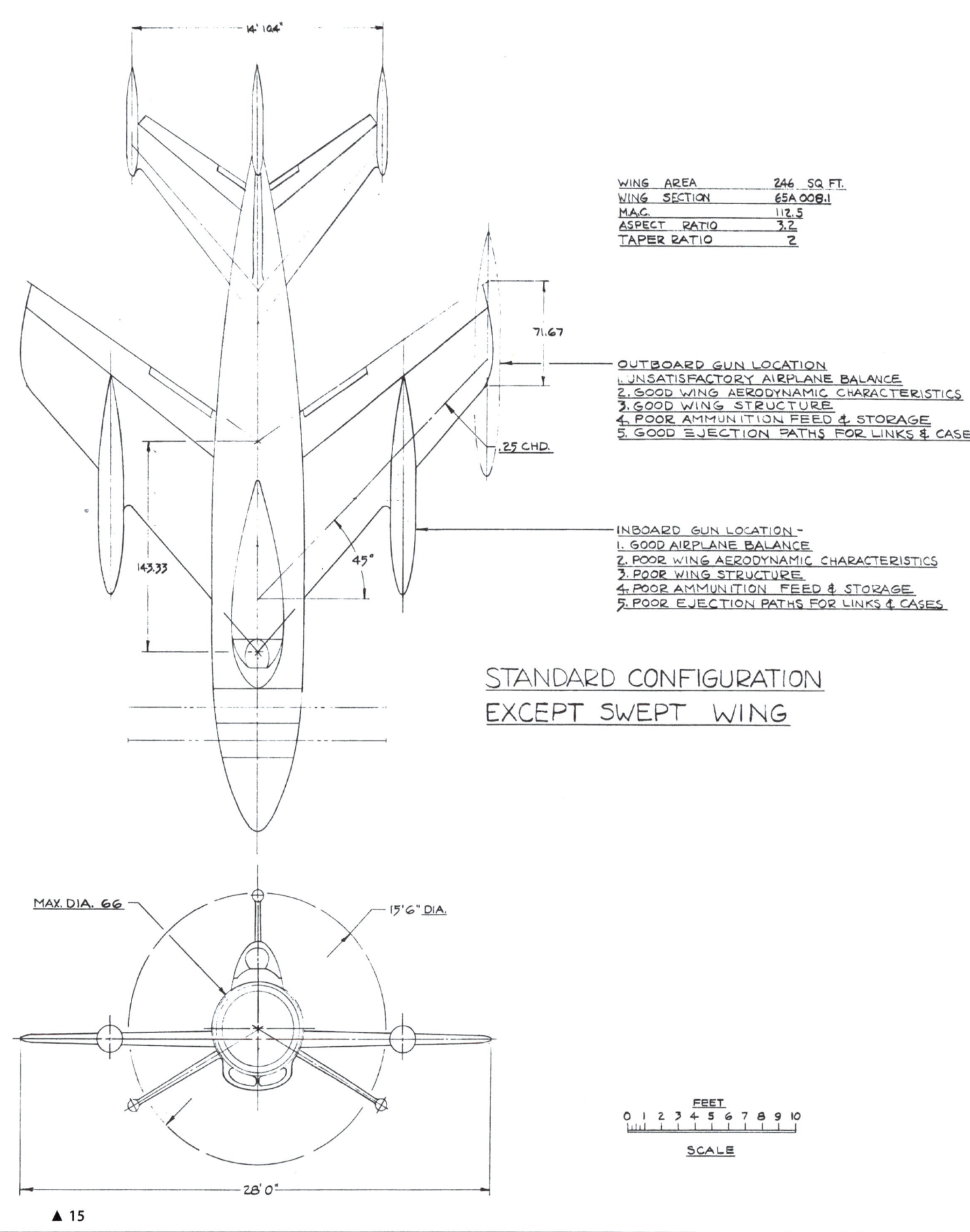

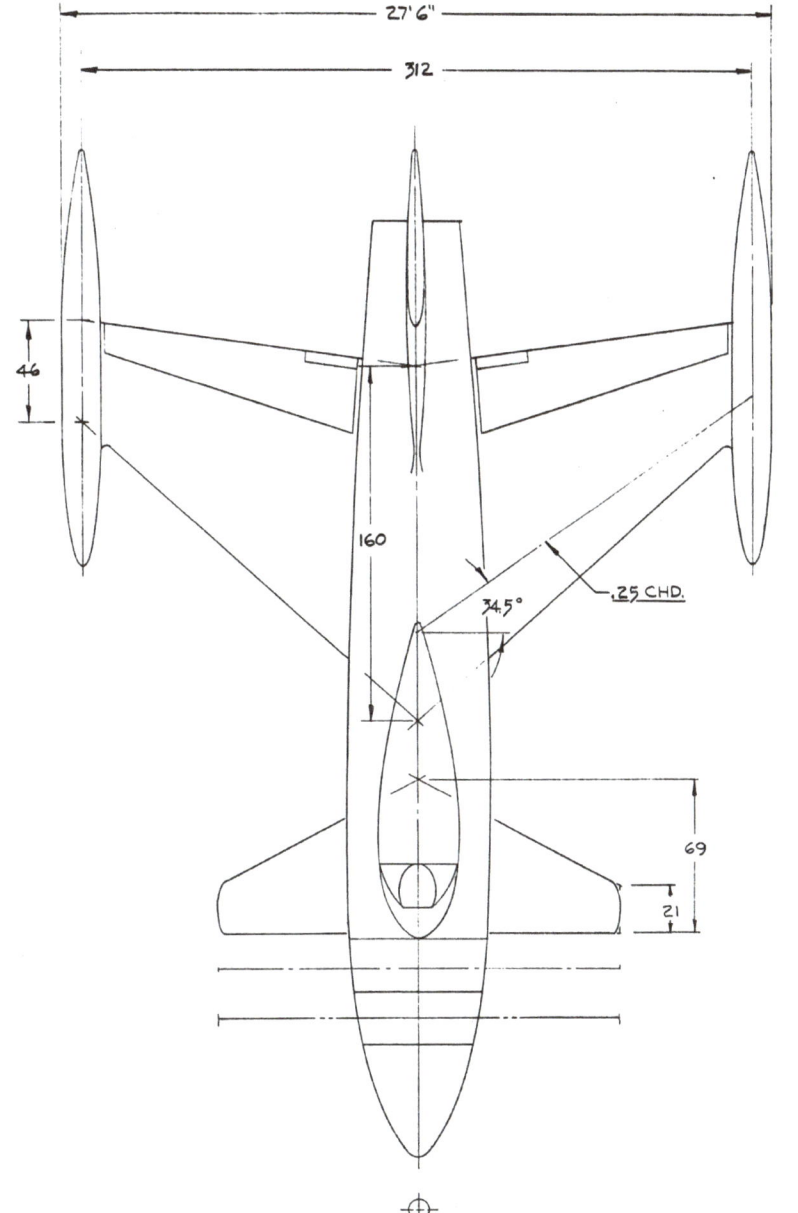

CANARD CONFIGURATION

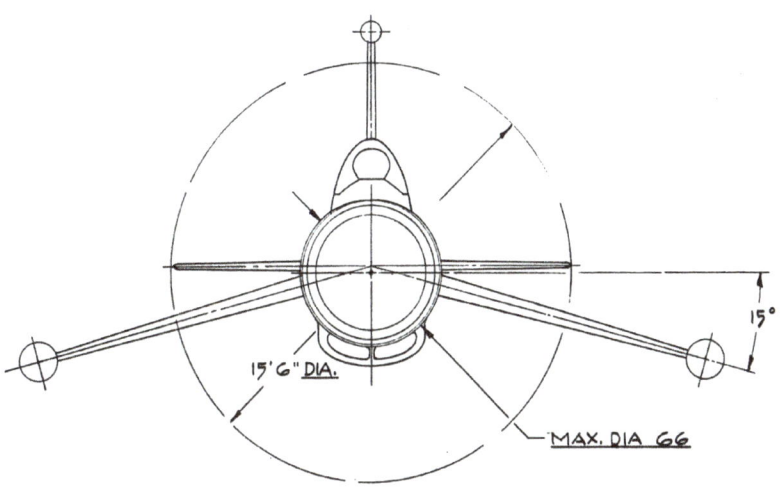

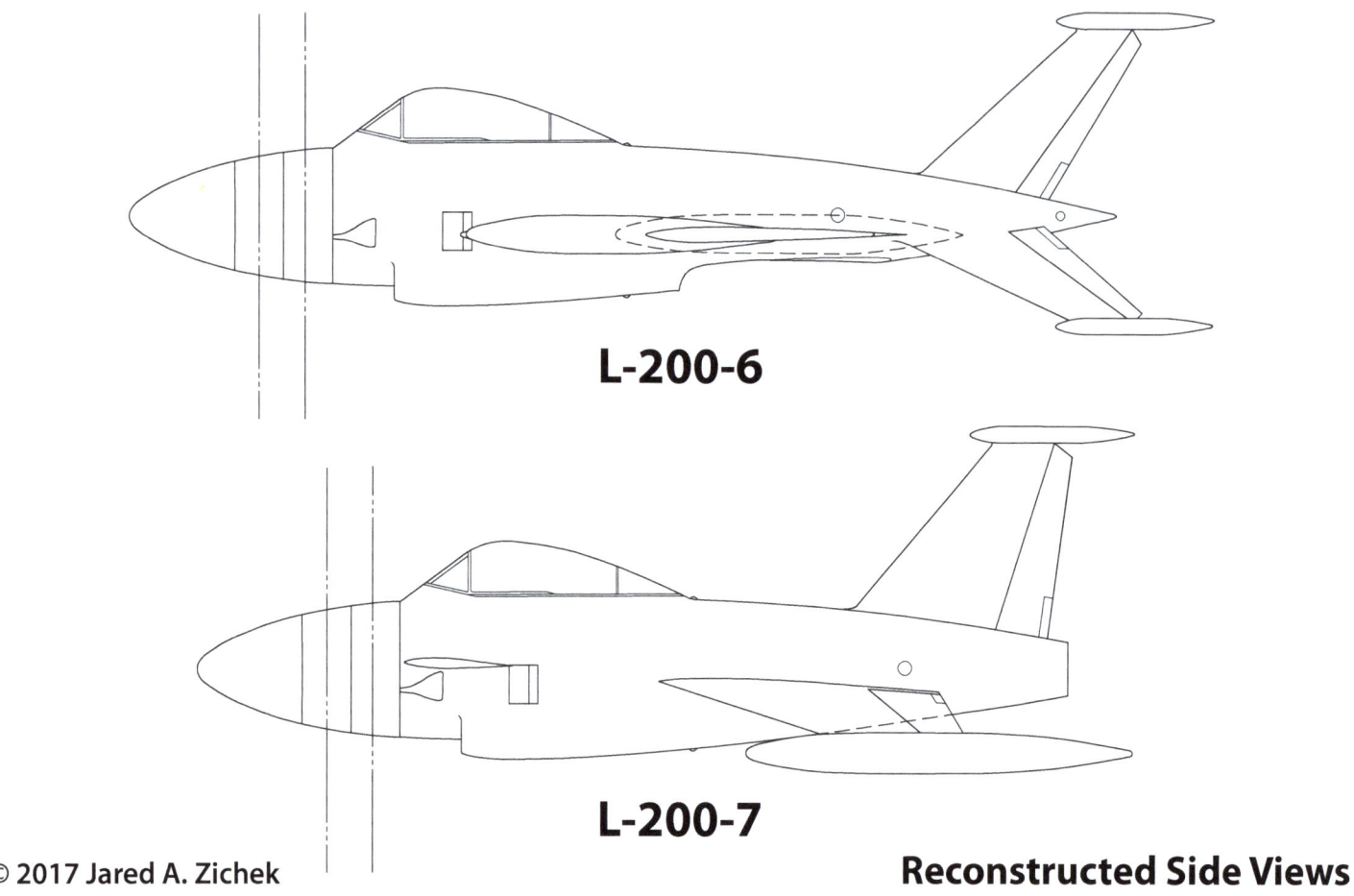

Reconstructed Side Views

© 2017 Jared A. Zichek

▲ 17

hoisting the airplane directly. By handling the airplane in a stiff-legged apparatus, it was possible to keep the airplane securely attached to its dock so that it did not sway with the roll of the ship. Furthermore, by thus handling the airplane, the entire service process could be done directly over the landing platform, taking up no additional space on the ship. The alternative to this type of a maintenance rig was to leave the airplane in its vertical position for all servicing and maintenance. Disadvantages of the vertical positioning of the airplane which were obviated by the hoisting system were:

1. The airplane was more susceptible to ice collection and weather in its vertical attitude.
2. Entrance of the pilot to the cockpit was complicated because of the lean-forward position of the seat while the airplane was vertical, leaving little space for the pilot to crawl into the seat; therefore, and additional seat adjustment had to be made.
3. Workmen were high up during maintenance operations and the motion of the ship would be aggravating unless the platform stabilization devices were operated during maintenance work.
4. Airplane could land in an oblique attitude which was not satisfactory from

Previous Spread

15) The L-200-6 was an alternate design with a swept wing; compared to the straight wing L-200-1, it was heavier and had inferior aileron effectiveness in hovering flight.

16) The L-200-7 canard configuration was also judged inferior to the L-200-1; in wind tunnel tests, it suffered from an extreme diving moment at a high angle of attack due to the aft position of the wing compared to the propeller.

This Page

17) Lockheed provided no side views of the L-200-6 and -7 in its report; to aid modelers in scratchbuilding these interesting designs, I reconstructed these side views, which are to the same scale as the views on pp. 16-17. In drawing these, I assumed they had the same basic fuselage design as the L-200-1, as shown in the plan on p. 14.

a maintenance or takeoff standpoint. This probably required a hoisting device for readjustment.

Some advantages existed, however, with the vertical maintenance scheme and these were worthy of further study:

1. It was simpler than the previously described rig if it did not have to hoist the airplane and did not require a mechanism for rotating the airplane.
2. The position of the airplane during storage was such that it was immediately available for takeoff in the event of a rapid scramble.
3. All sides of the airplane fuselage and both sides of the wing were equally accessible for maintenance operations since the airplane was supported by its natural landing gear on the tail.

It was suggested that one of the major contributions which a prototype program could make was in the development of the servicing and maintenance equipment. The design of the fuel system, oil system, and ammunition-carrying system was such that loading and replenishing the airplane could be done either horizontally or vertically.

Lockheed designed a cart which served as a maintenance or moving stand when the airplane was away from its landing platform. This rig could be carried aboard ship and attached while the airplane was suspended above the landing platform, then hoisted overboard with it when unloading.

Aerodynamics Summary

The discussion of performance, stability, and control of this airplane was divided into two parts following the two contrasting flight regimes. The first regime included the transition from zero velocity to power-off stalling speed, and from power-off stalling speed to hovering flight, which constituted an aerodynamic study unique to this convoy fighter. The second regime, which included all the normal flight operations of the airplane above power-off stalling speed, followed conventional aerodynamic analysis for such high speed fighter aircraft. Three reports accompanied the proposal which carefully analyzed the transition performance, stability and control, and the performance and control characteristics in the normal flight regime. The following discussion presents a synopsis of these three aerodynamic studies.

Part I—Takeoff and Landing Transitions

The transition range considered was a range between zero flight speed and approximately 150 kts. Throughout this range, two aerodynamic effects were predominant— effects which usually required very little consideration. These effects were:

1. The characteristic forces and moments on the components of the airplane under conditions where air flow separation was the governing influence.
2. The effects which resulted from an abnormally large relation between propeller slipstream speed and normal free stream speed.

Although it was possible to obtain crude estimates of these effects by calculation, Lockheed considered it necessary to utilize wind tunnel results in order to feel confident that the desired transition performance and control could be realized. Wind tunnel tests were made, therefore, in the Lockheed Aerodynamics Laboratory.

The component aerodynamic forces which influenced the airplane at each particular relative angle of attack could be summed up as follows:

1. The propeller created both a thrust and cross wind force. The thrust gave rise to slipstream velocity. The cross wind force gave rise to a pitching moment, particularly at the extreme angles of attack approaching the hovering position, which was of major importance. These moments created the positive pitching effect required for stability in hovering flight which was opposite in nature to the conventional requirement for negative pitching moment in normal airplane flight.
2. The wing was utilized for its normal functions in normal airplane flight,

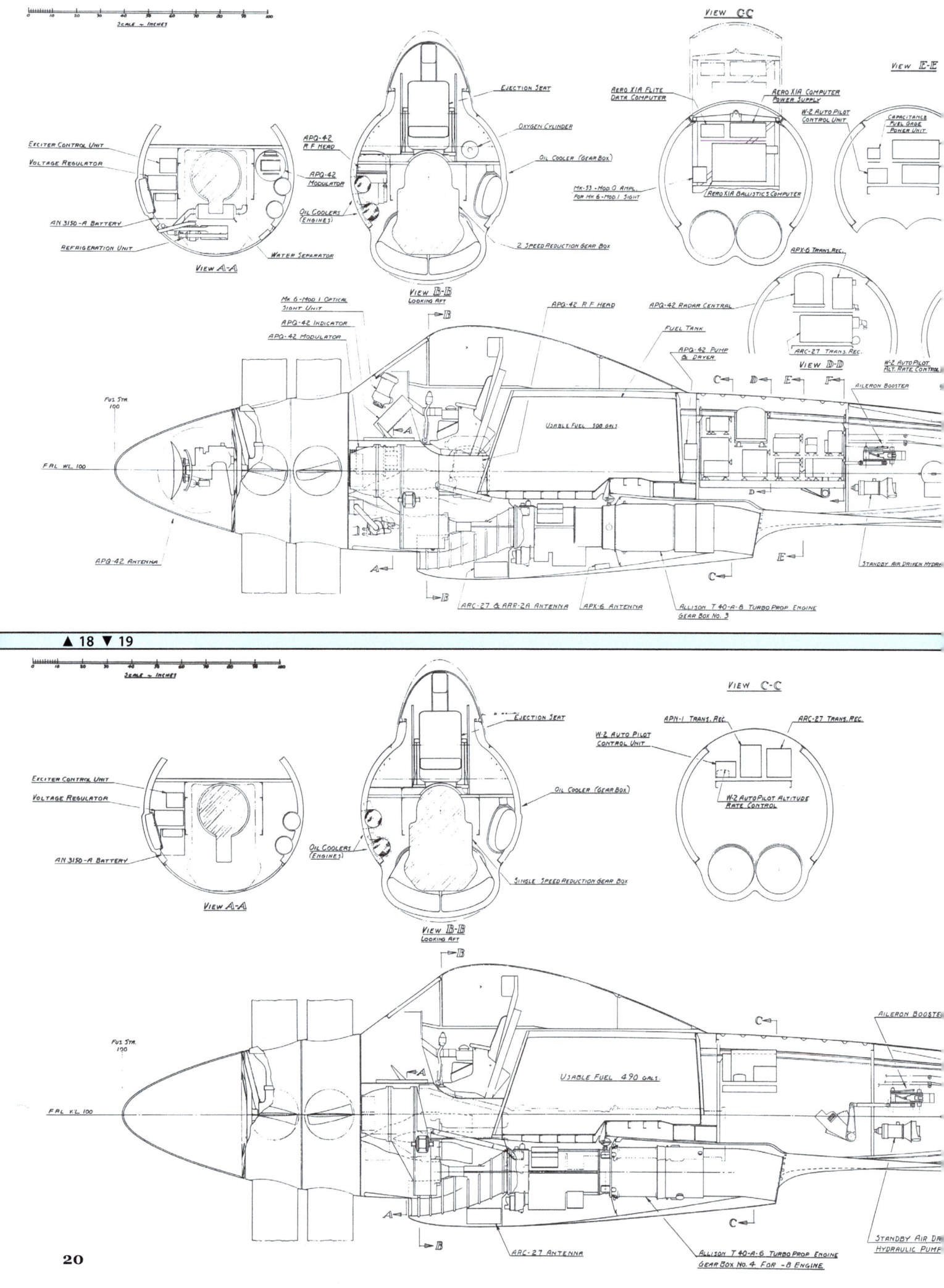

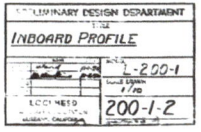
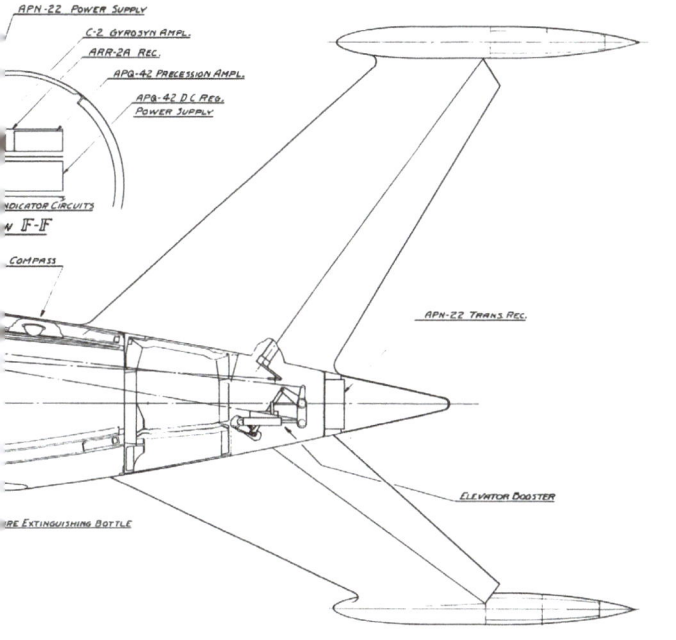
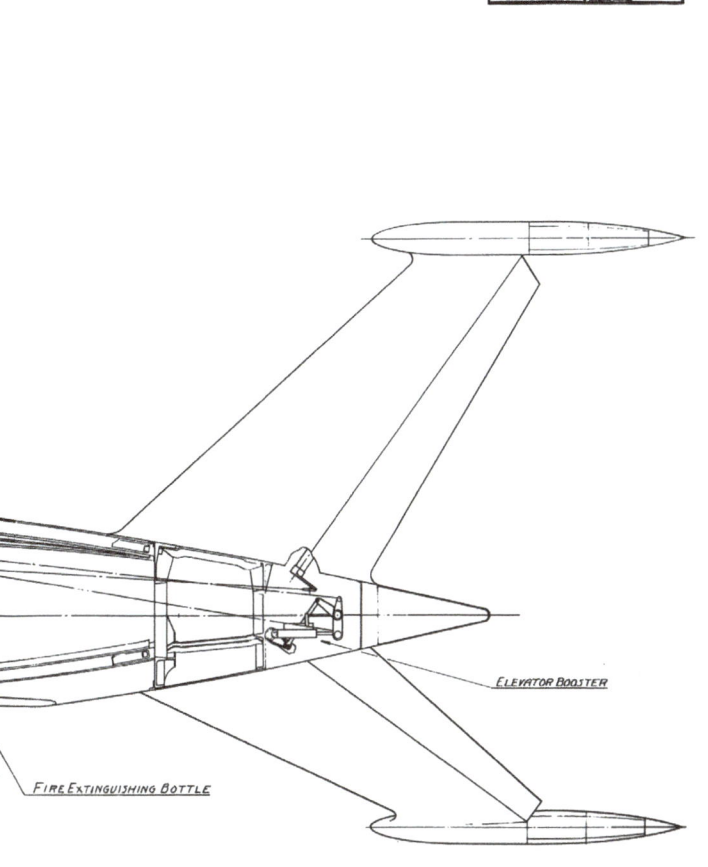

but at high angles of attack, wing lift due to slipstream acted in the drag direction. This slipstream effect on wing forces and moments, particularly at air speeds approaching zero, was a large factor in determining the ability of the airplane to follow a desired transition procedure.

3. The tail surface operated in a conventional manner in producing diving moments. The important effect of slipstream on the tail surface was to create sufficient dynamic pressure such that pitch and yaw control would operate at extremely low translational velocities. The problem of tail surface control effectiveness became one of ensuring that the slipstream influence for the high angle of attack range was sufficiently strong such that control was maintained up to higher translational velocities than would be required in any transitional maneuver.

4. Roll control was required to allow for propeller torque inequalities and gust effects, particularly in the hovering regime, and for sufficient control in the normal sense so that the pilot could execute his desired maneuvers. Although it appeared possible to use propeller controls for roll control, it was not deemed desirable, and normal aileron type controls were included in this design. Thus, the slipstream effect had to be sufficient to a relatively high translation speed to ensure no lack of roll control at the speeds required throughout the transitions.

18) Inboard profile of the production Lockheed L-200-1 showing the general arrangement of the equipment, power plant, cockpit and internal structure. This version was powered by the Allison T-40-A-8 turboprop engine.

19) Inboard profile of the Lockheed L-200-2 full-scale prototype powered by the less powerful Allison T-40-A-6, an alternative to the 0.766 scale demonstrator discussed in the second volume of this book.

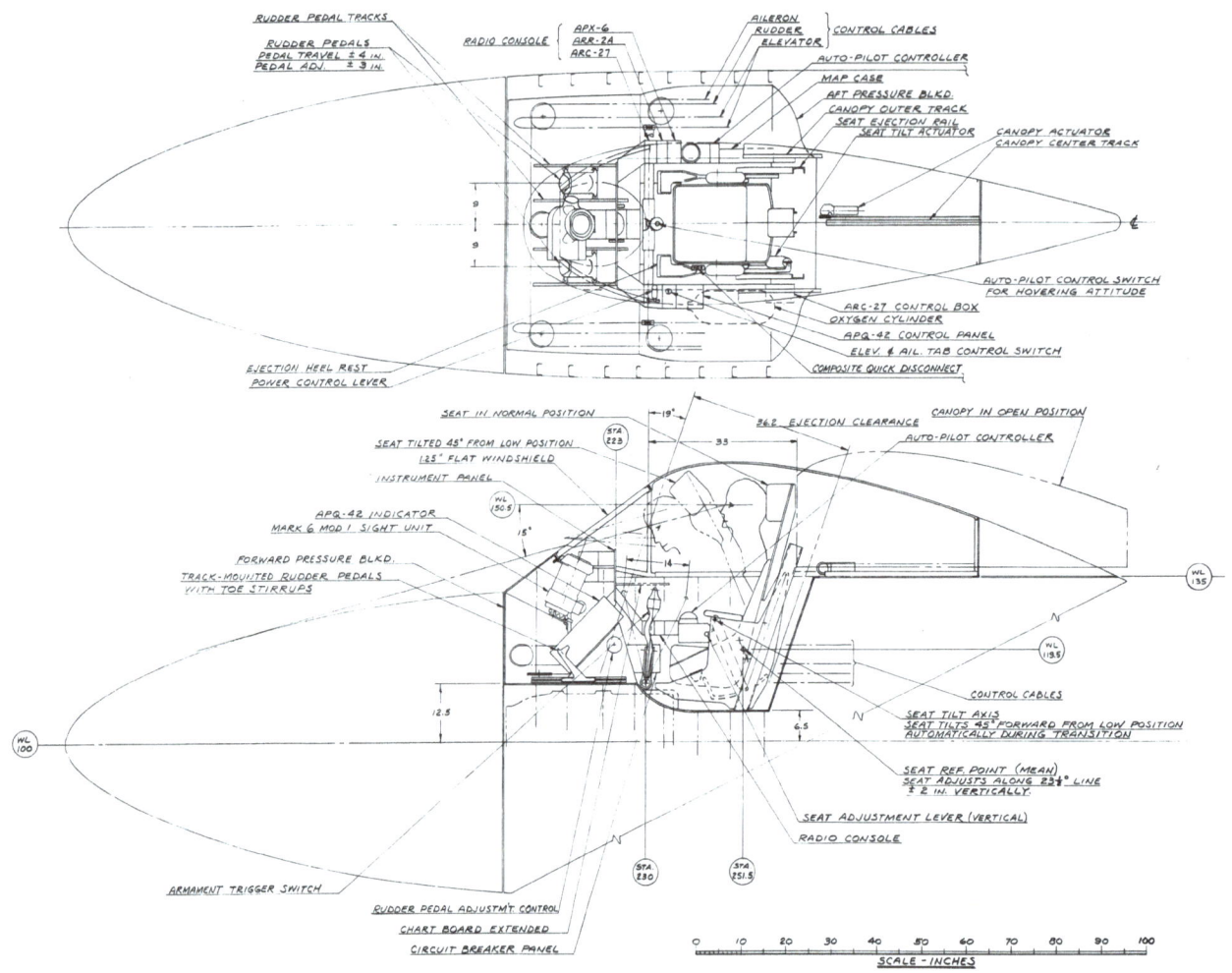

▲ 20

Transition Control Devices

The airplane incorporated conventional types of aerodynamic surfaces with the possible exception of the use of full-span ailerons and flaps. For the normal flight regime, the pilot's controls were the standard stick and rudder type plus the modern interceptor automatic pilot system. For flight in the transition range, the pilot's controls included one with which he could request angular pitch rates, plus one with which he could request absolute pitch and yaw angles for the range close to the hovering attitude. In addition, he had the ability to request a power variation which was automatically controlled to maintain constant altitude. Roll stabilization was automatic such that he could be assured of zero rate of roll. For fore and aft horizontal translation from the hovering position, which was required to allow the pilot to position and hold the airplane over the landing platform, the aileron could be deflected as flaps. This was desirable to maintain airplane angular attitudes within 8° of vertical for the highest steady state horizontal translational speeds required for landing.

Takeoff Transition

After vertical takeoff, the capabilities of the airplane with full power were such that the pilot could choose a constant pitch rate on the order of 5° to 10° per second as he desired. It was recommended from this study that a 5° per second pitch rate be used which resulted in transition in 21 seconds using 940 ft of altitude. The choice of pitch rate was based on the optimum change of altitude and on the desire to obtain normal level flight airplane speeds. The constant pitch rate system was chosen both because it was the simplest system to provide in the airplane and because it resulted in a favorable speed, altitude, and acceleration combination. In this

20) The cockpit arrangement of the L-200 was relatively conventional in spite of the unconventional requirements placed on the pilot. The most noteworthy feature was the seat, which was arranged to pivot about an axis parallel to the wing axis in order to raise the pilot's head as the airplane approached its vertical attitude.

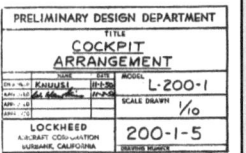

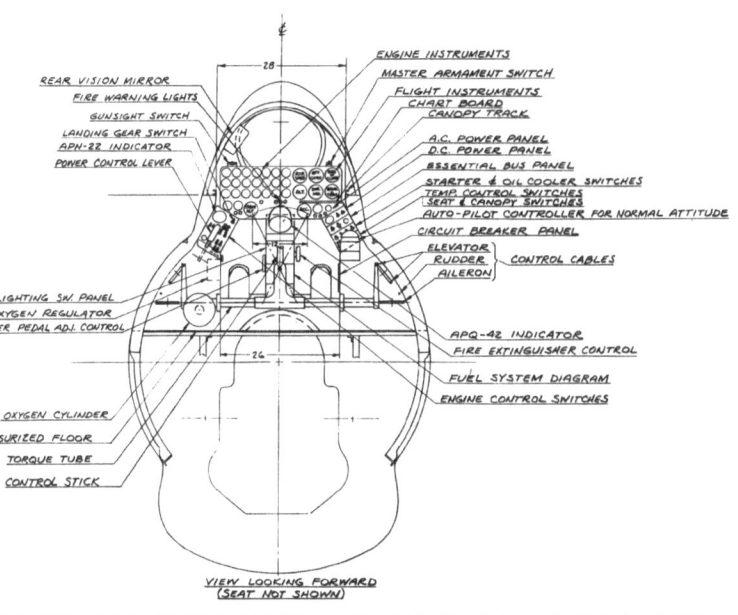

takeoff transition, the effective airplane angles of attack never exceeded those used in normal airplane flight, and controllability was not a difficult problem. However, the nature of the changes in control position with speed was such that an initial deflection created the maneuver and, subsequently, a control position had to be maintained to keep from increasing above the desired angular pitch rate. It appeared possible, after practice, for the pilot to accomplish this maneuver without the automatic devices, but he had to be limited in the maximum possible pitch rate such that he did not arrive at the horizontal attitude with insufficient altitude and speed. The typical transition picture shown on p. 11 was determined from actual wind tunnel test information.

Slowdown Transition

The problem of slowing down below power-off stalling speed and attaining hovering flight was solved by a method very similar to the takeoff transition in that it was based upon the use of pilot chosen pitch rates. However, it appeared that the critical and most desirable slowdown transition was one that was made at constant altitude, and calculations based on wind tunnel tests demonstrated that this was possible. It was assumed in the calculations that the automatic power control maintained the thrust required to ensure constant altitude. This thrust variation had been shown to be one in which the transition was initiated from a speed of approximately 150 kts at an angle of 15° and with relatively low engine power, and thereafter power was automatically increased to a maximum of approximately 70% as hovering altitude was approached. For the critical heavyweight conditions, the control movements were such that the elevator was moved to initiate the pitch rate, and sometime thereafter had to be reversed to ensure that the pitch rates did not become excessive. A 5° per second pitch rate transition was considered optimum in that it required 15 seconds, and resulted in a translational velocity when the airplane attitude had reached the vertical of 31.4 kts. It appeared that the pilot could make this slowdown transition manually, although it was probable that he would make some gain in altitude due to his logical manipulation of the power control. Care had to be taken to limit the maximum pitch rate so that the airplane did not reach the vertical attitude with too high a translational speed. The wind tunnel had demonstrated that there was a critical top speed for traveling horizontally in the vertical attitude at which the pitch control deteriorated. However, from the slowdown transition calculations there appeared to be adequate margin below this critical speed. Aileron roll control also had a critical maximum speed which appeared sufficiently high such that the airplane would not run out of control effectiveness in any normal transition maneuver.

Descent and Landing

The airplane was capable of hovering at altitudes up to a maximum of approximately 19,600 ft. Wind tunnel tests showed that rates of vertical descent up to values of 2,000 ft per minute were entirely satisfactory from considerations of control. The constant altitude power control could also be used by the pilot to request rates of descent such that he could ensure not exceeding the critical descent speed. Wind tunnel tests had not been run to determine the maximum descent rate because a 2,000 ft/minute value appeared adequately high.

When the airplane reached the landing altitude it was necessary to translate horizontally, maintaining constant altitude to position the airplane above the landing net. The illustration on p. 13 demonstrates the airplane attitude required to translate both in the fore and aft and in the side direction as derived from the wind tunnel test data. The pilot executed his desired translation by setting the angle of the longitudinal axis at the desired amount from the vertical with the constant altitude control maintaining the power required. The airplane accelerated horizontally towards its steady velocity and it was easily checked by the pilot by a rapid change in airplane attitude.

Very little angular change in yaw from vertical was required to attain desired velocities in side translation. Fore and aft translation, however, normally required sizable pitch angles from vertical to translate at equivalent speeds, due to the influence of the forces acting on the wing. Therefore, the ailerons were deflected as flaps for rapid fore and aft horizontal translation. These flaps were very effective in reducing the required pitch angle from vertical in the desired horizontal speed range.

Hovering Stability

Preliminary dynamic stability calculations for the hovering attitude indicated that at high weights, an airplane disturbance from vertical would be counteracted by the inherent static stability, but that the dynamic characteristics would be unstable. Conversely at low weights, which meant lower engine power, the results of a disturbance from hovering flight tended to be increased by an unstable static stability, but with dynamic subsidence about the static variation.

21) Blueprint of the L-200-1 airplane final power plant installation; the basic power sections were those of the Allison T-40A-8.

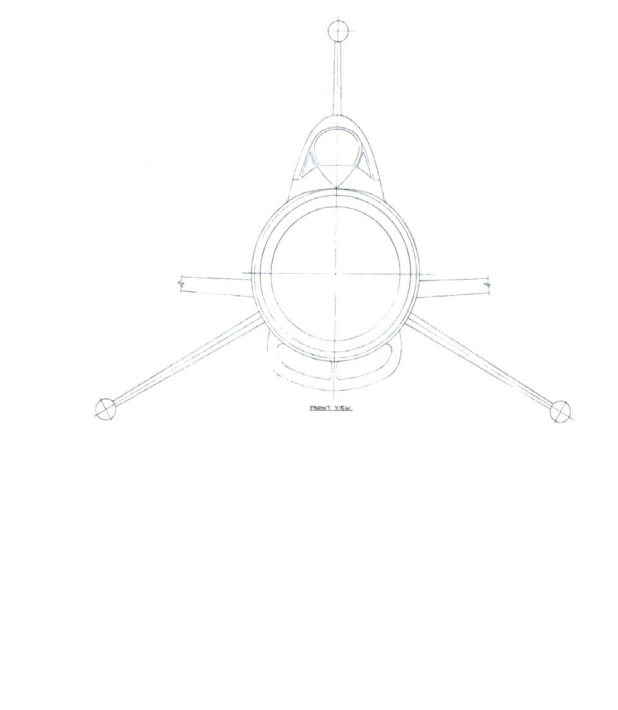

▲ 21

However, the importance of these natural stability characteristics of the airplane in hovering flight had to be judged relative to the dynamic oscillatory period and to the divergence time. Fortunately, calculations indicated relatively long periods and times to increase amplitude as compared to the normal pilot's reaction time. This meant that without the automatic stabilization devices, the pilot could hover the airplane satisfactorily. These facts had been calculated theoretically, and had also been indicated by experience with a flight model of the airplane.

Part II—Normal Flight Characteristics

The complete transition analyses were based upon L-200-1 airplane because, relative to the transition regime, there was little difference between the aircraft presented. However, in this section, the normal flight characteristics of high speed, climb, and range were discussed and, in these respects, the various airplane proposals differed.

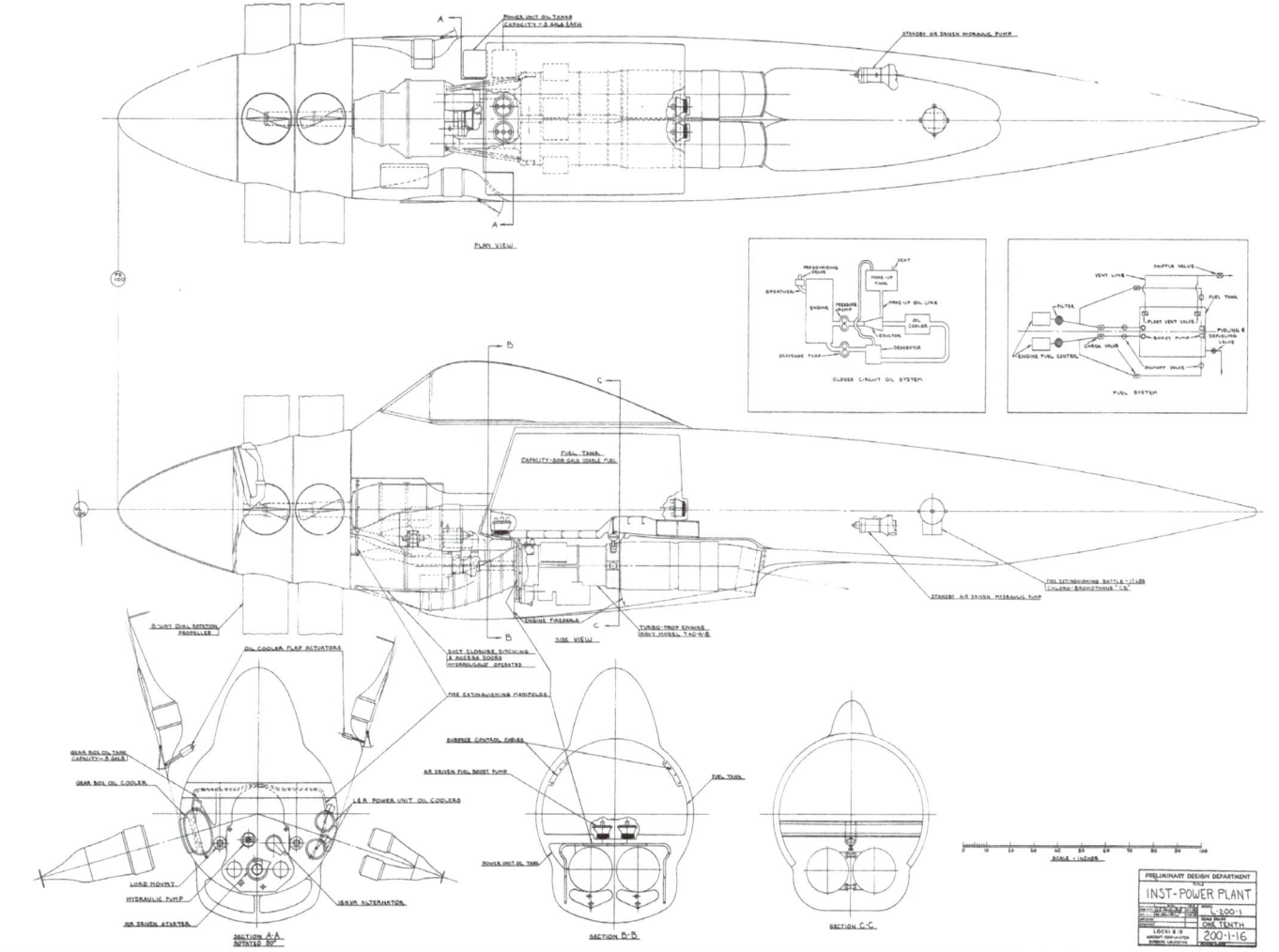

Propeller Selection

Information available from the propeller manufacturers showed that the largest propeller was desirable in that increased propeller thrust more than compensated for increased propeller weight. Therefore, the objective of propeller selection was to choose the maximum thrust propeller which, for the basic airplane L-200-1, was the eight-bladed dual rotating 15½ ft diameter type. Transition analysis only required that the maximum possible thrust be available. The increased propeller cross wind forces which increased propeller solidity would not adversely effect the transition performance. The optimum choice envisaged the use of this propeller with a two-speed gear ratio to obtain the best thrust for transition and the best efficiency for climb and high speed flight.

The choice of a single versus dual gear ratio determined the added weight and complication in the airplane. The prototype L-200-2 and L-200-5 airplanes had single propeller gear ratios chosen as optimum for the transition regime. For the production airplane, there was a question which had to be considered further. The proposal requirements could be satisfied with either a two-speed gear ratio or a single speed system plus an engine afterburner. A two-speed gear ratio and an afterburner improved the high speed and climb performance. The thrust available varied depending on gear ratio and afterburner installation maintaining the optimum thrust characteristics for transition with a propeller rpm of 750 and 1100.

An entirely different approach could be taken with a single speed propeller whereby one of the dual units could be put into high pitch with essentially zero thrust and the other propeller allowed to absorb all the power; however, tip speed corrections appeared to be excessive for this type of operation.

All configurations were able to meet the high speed requirement except possibly the last which could not be truly evaluated with the test data available.

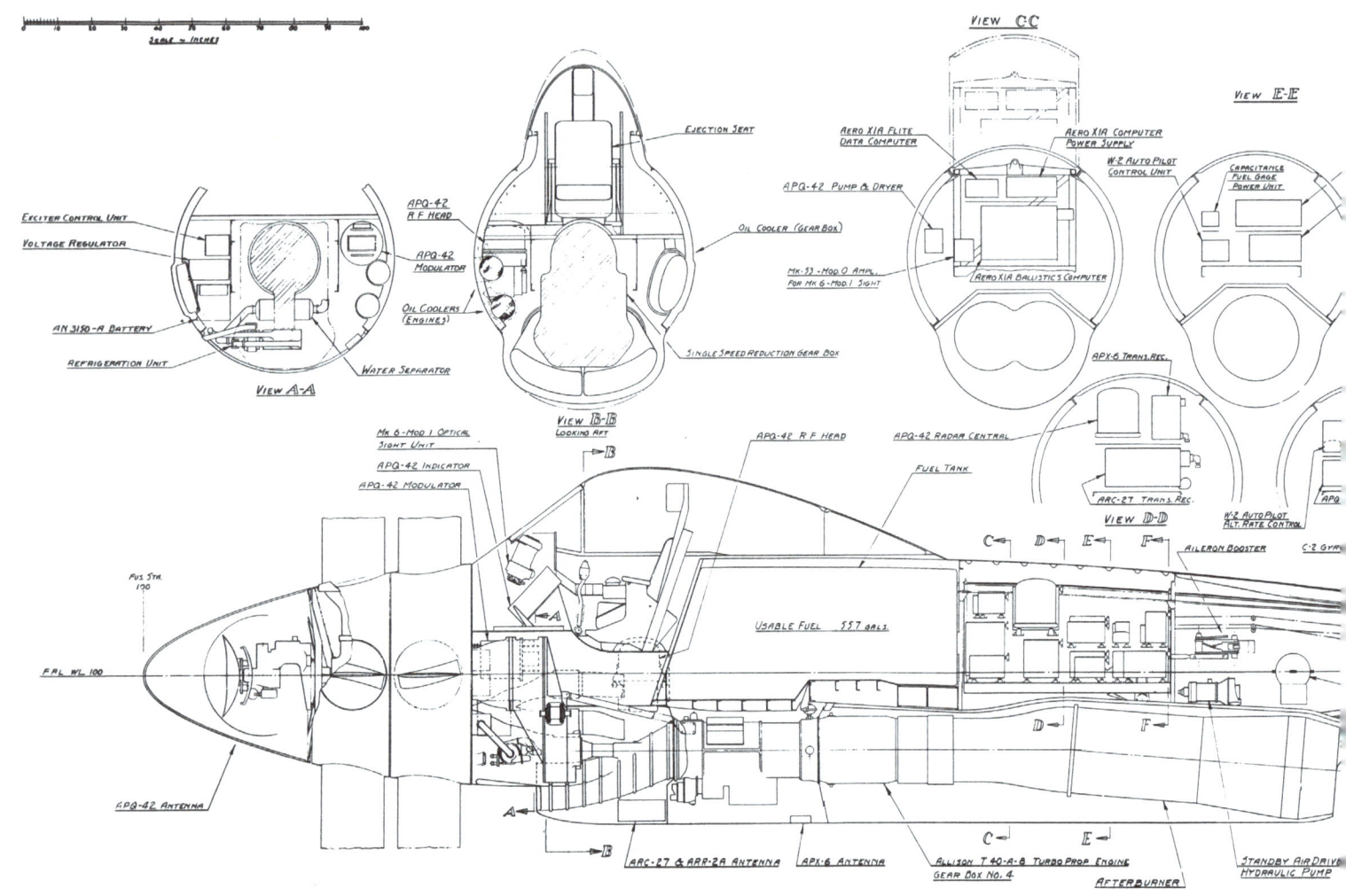

L-200 Tactical Airplane Performance

The L-200-1 exceeded the Navy requirements for high speed and combat ceiling, fulfilled the complete desires for the basic radius mission and came close to obtaining the required time to climb to 35,000 ft. Since the drag on this airplane was low and the calculated propulsive efficiency was as high as could reasonably be expected (88%), the possibility of improving this climb time was doubtful. Optimistic assumptions may have been made on the DR-72 airplane which accompanied the Navy's proposal material and permitted the time shown in that report. Since NACA tests had proven a poorer "e" for a higher swept wing as compared to an unswept wing, and the proposed airplane was lighter in weight, the time to climb of the L-200-1 should have been slightly superior to that of the DR-72. Neither airplane met the requirement unless propulsive efficiencies were considerably superior.

Prototype and Alternate Airplane Performance

The performance of two prototype airplanes was studied, the first being a 0.766 "scale" model of the Convoy Fighter powered by a Double Mamba turboprop engine, and the second a full-scale reduced weight airplane with an earlier Allison XT-40A-6 engine. Both airplanes exceeded the small amount of performance requirements of NAVAER OS-121. However, the performance capabilities of the L-200-2 (full-scale prototype) were well above those for the 0.766 scale model. This model would also provide full-scale characteristics of the tactical airplane since no disproportionate cockpit size and incorrect fuselage interference effects existed, thus providing an answer to the full-scale flight characteris-

22) Lockheed studied the option of adding an afterburner to the Allison T-40A-8 to eliminate the necessity of a two-speed gear box. While it increased the top speed of the aircraft, the afterburner also increased its overall weight. The company recommended additional study to determine whether the high speed advantage was worth this penalty.

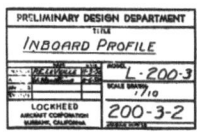
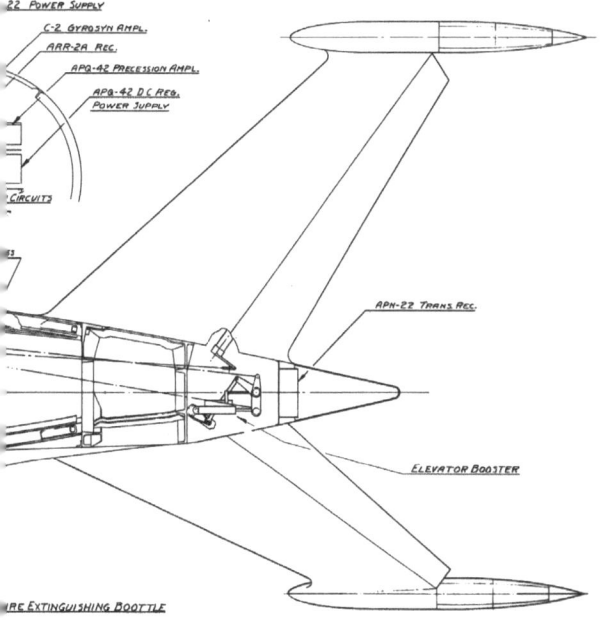

Stability and Control

The stability and control of the L-200-1 in the normal airplane flight regime fulfilled most of the requirements of NAVAER Specification SR-119-B. In many ways that specification was not applicable to an airplane which could fly and maintain altitude to zero velocity. The requirements for normal flight stability and for stick force variation in high speed maneuvers were satisfactory. A particular condition of SR-119-B which was obviously difficult to fulfill concerned stalling characteristics. Since this airplane incorporated relatively greater thrust than any of its predecessors and, since it entered regimes in which the outer wing sections were completely stalled, it was obvious that control manipulation was required of the pilot when operating without automatic pilot devices. However, the reliance on automatic stabilization devices was becoming more acceptable today because they appeared necessary even for relatively conventional aircraft. Since this design envisaged identical types of automatic stabilization, the use of these devices to some degree in the satisfaction of requirements of SR-119-B was acceptable.

The most marked difference relative to stability of this airplane was the usage of an extremely far forward center of gravity. This was required to compensate for the large destabilization effect of the propeller, but fortunately was in the proper direction for the optimum configuration for transition performance. In addition, the deletion of any normal landing requirement, which always created difficulties of elevator control due to ground effects, made the usage of a far forward center of gravity feasible.

Alternate Design Discussion

The study leading up to this proposal was started by a careful comparison of the design submitted with the Navy proposal material as compared to alternate designs. The use of a coaxial propeller in the conventional propeller position was largely dictated by the requirements of minimum all-up gross weight. With the existing limitations of engine thrust-weight ratios, it appeared undesirable to deviate from this engine-propeller-fuselage relationship. The most promising configurations were the normal swept and straight wing types, plus the tail-first canard configuration. The choice of the straight over the swept wing was largely a matter of weight, but there was one aerodynamic advantage which recommended straight wing configuration. Wind tunnel tests indicated that the aileron effectiveness on the swept wing was very poor and that extremely unconventional means would be required to obtain roll control with this swept configuration in the hovering range. Aileron effectiveness on the straight wing configuration appeared entirely satisfactory throughout all the speed and angle ranges which were required.

The canard configuration appeared to have advantages relative to the design provisions for takeoff and landing. Aerodynamically, the canard configuration appeared questionable and subsequent wind tunnel tests had shown that this configuration would be unsatisfactory. The greatest aerodynamic difficulty of the canard was the extreme diving moment at high angle of at-

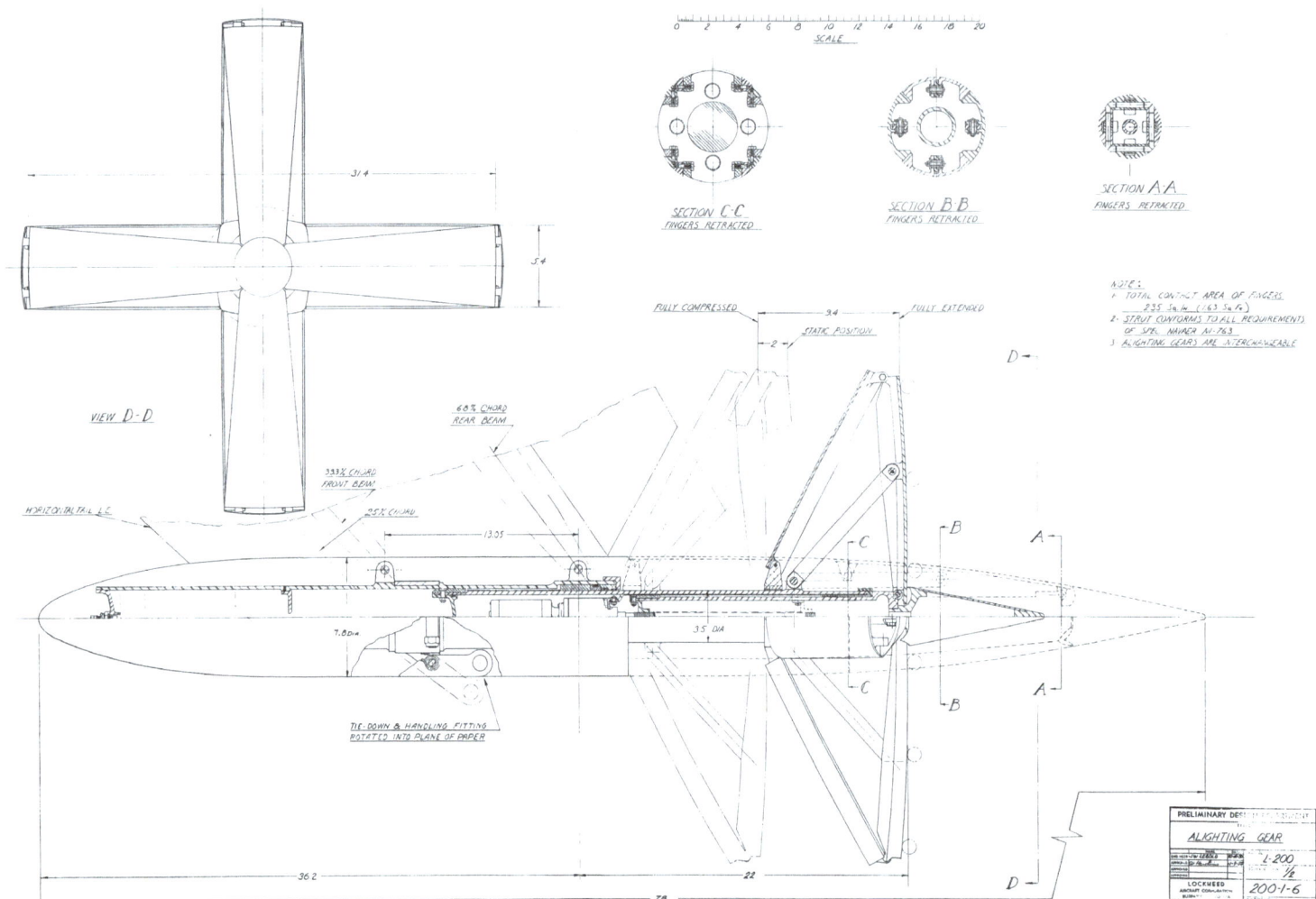

▲ 23

tack due to the aft position of the wing compared to the propeller. Without a considerable weight increase to lengthen the fuselage, it appeared that this problem was insurmountable. Concurrent with this, the effectiveness of the nose trimmer for longitudinal control appeared too small due to the reduced slipstream velocity in the position close to the propeller.

Development Program

Considerable study was made of the various methods of developing an airplane of the L-200 type. Obviously, a very complete research and mock-up program was required to guarantee the ability to make vertical ascents and descents under complete control. Also, a great deal of work was required to obtain the optimum transition between vertical and horizontal flight. Lockheed concurred with the Navy in that it was extremely important to flight test a prototype airplane as soon as possible. Having designed the Mamba-powered prototype, suggested by the Navy, and compared it to the full-scale L-200-1, it was immediately apparent to Lockheed that the sizes of these two airplanes were so nearly the same that there would be very little time saved in making the Mamba-powered airplane against constructing a stripped full-scale L-200-1. In comparing the development trend of the Allison T-40 engine, studies seemed to indicate that it was entirely feasible to fly a prototype L-200-1 airplane in the same time that one could fly the Mamba-powered airplane. The advantages of this procedure are outlined below:

1. The method of gearing the power sections to propellers on the T-40 was preferred because it permitted both propellers to operate from one power section, increasing the available thrust during one engine power-off operation which could save the pro-

> **23)** When landing on a ship, Lockheed's Convoy Fighter was designed to touch down on a landing net with its three-pronged empennage. This drawing of a typical tail tip pod shows the mechanism for retracting the landing pads and obtaining adequate shock absorption.

totype airplane. This was not possible with the Mamba engine without a complete gearbox development.

2. Having the prototype airplane complete to size and identical to the eventual tactical airplane permitted development of valid aerodynamic data beyond the development of the landing and takeoff procedures.

3. Having the prototype identical in shape and size to the tactical airplane meant that controls, cockpit arrangements, booster system, hydraulic system, and electrical system would all be essentially identical to the final airplane; any development problems that arose in prototype operation would essentially be solved and would not recur in the flight of the tactical airplane.

4. Ground handling equipment, including landing facilities and maintenance facilities, would be identical between the prototype and first tactical airplane, and the development would not have to be duplicated.

5. The major detail design problem aside from the landing and takeoff was the problem of maintaining sufficiently low weight so that a usable margin between thrust available and required would be realized on the final design. With this in mind, it was proposed that the full-scale prototypes be used for a) measuring loads in flight and b) that an identical article be used as the static test airplane for the tactical version in spite of the fact that its design gross weight would be considerably lower than the tactical airplane. This required more modifications during the static test program but would result in a substantially lighter final aircraft.

6. Since it appeared that modifications were required on the Mamba power plant for vertical flight, and, since it appeared entirely feasible to fly a full-scale T-40 prototype without the two-speed modification to the gear box, the time for developing either prototype could be approximately the same. From a cost standpoint, it was estimated that developing full-scale prototypes in Part I of the development program would be far cheaper in arriving at a tactical airplane than if a Mamba type prototype was developed. This was due to the fact that only three articles needed to be manufactured instead of five and the transition from prototype to tactical airplane would be much simpler. This did not imply that the prototype part of the suggested program itself would be any more expensive since the development of any airplane of this type appeared to involve just as many research problems, whether it weighed 7,500 lbs or 12,500 lbs.

A possible disadvantage of the suggested procedure concerned the reliability of the T-40 engine compared to that of the Mamba. At the time, the Mamba was probably more reliable, but considered from a view of 18 months additional time, and also the advantage of the gearing method of the T-40 type engine (allowing flight on one engine), it seemed that satisfactory reliability should be available from the T-40A-6 engine by the time it was required for a prototype airplane. It was necessary to ensure proper delivery of the propeller gearbox to meet the required time schedules. The Allison Corporation provided the following dates applied for the various configurations of their gear boxes:

No.	Type	Months
1	Standard gearbox revised for vertical flight	15
2	Standard gearbox plus two propeller speeds	24
3	Large offset gearbox plus two propeller speeds	24
4	Large offset gear box single speed	21

Note: Front accessories, if desired, took an additional 3 months on Numbers 3 and 4.

The prototype L-200-1 airplane could be flown with the standard gear box revised for vertical flight which would be available in

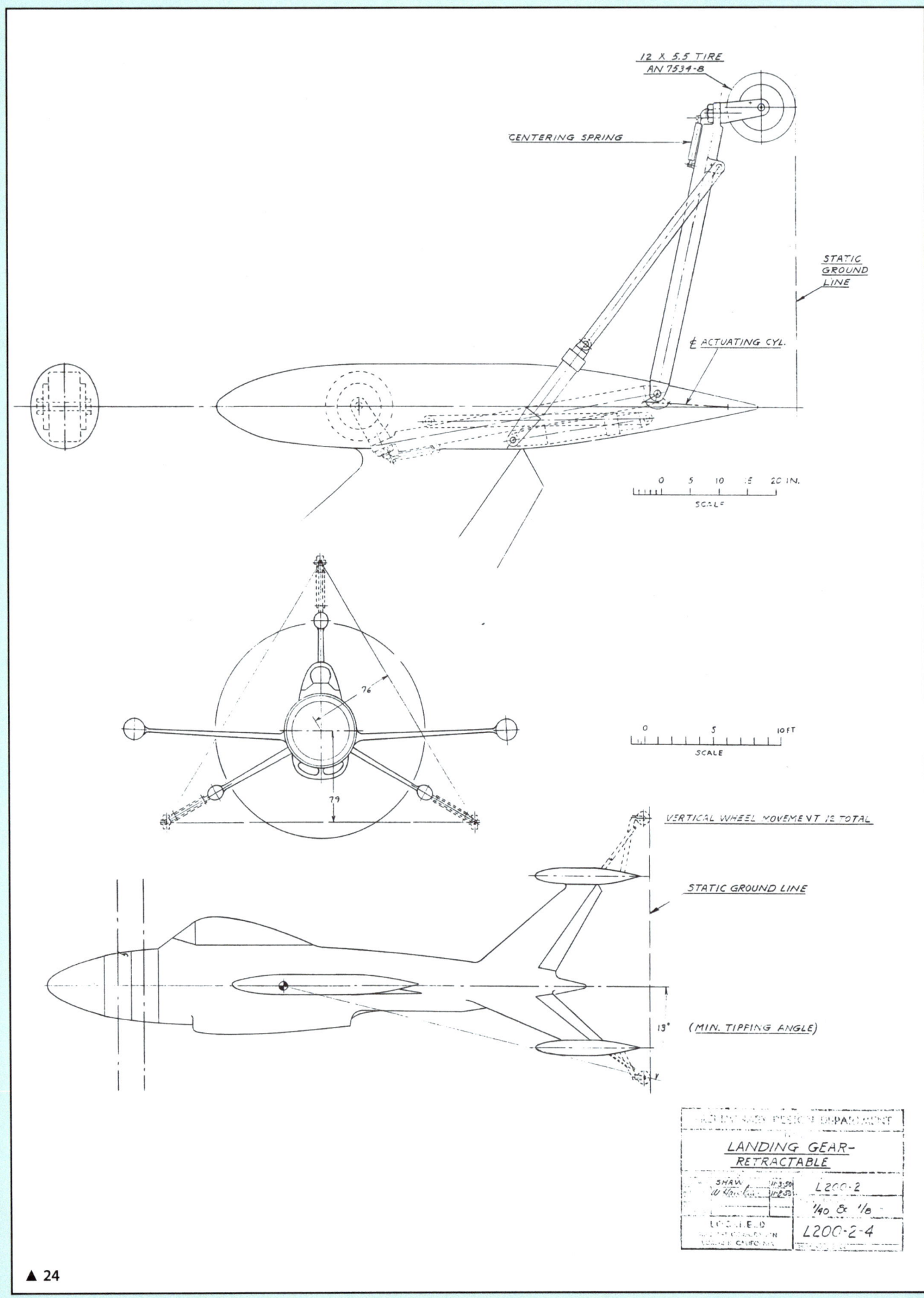

ARMAMENT INSTALLATION

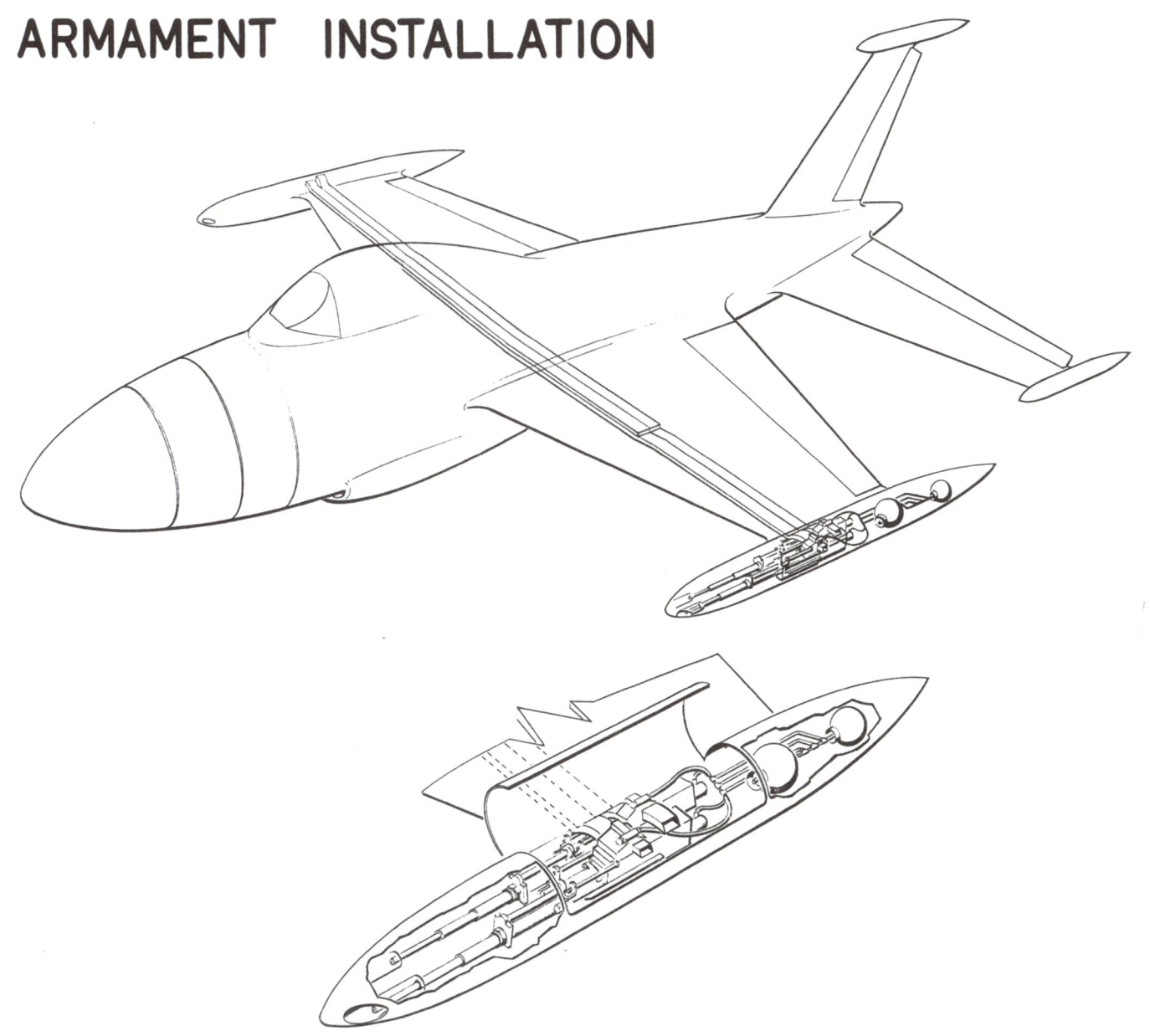

▲ 25

15 months. This procedure was not desirable, however, since the high position of the engine changed the fuselage structure, caused very poor duct inlets, broke the wing structure and lengthened the tail pipes. It was considerably more desirable to obtain Gear Box No. 3 or No. 4 in fifteen months or delay the program enough to obtain these units. With the above development spans, the production airplane deliveries would be delayed six months for Gear Box No. 4 and nine months with Gear Box No. 3. Even with the maximum gear box delay, a production airplane was available at the same time as with Program A. In spite of this, every effort was made to get Gear Box No. 4, or preferably 3, developed in 15 months.

The reasons for proposing such a prototype program could be best obtained from consideration of Program A and Program B. Program A indicated the proposed Navy program as regarded time for various phases of development. Lockheed believed it reasonable to make tunnel tests, preliminary design, and the required research within a 6 month period. In an additional 12 months, the first of the two Mamba

24) Lockheed studied this alternate retractable wheeled landing gear as an interim installation which could be used for the first vertical flights of the airplane or prototype while developing the simpler tactical system. The actual XFV-1 was equipped with a less complex fixed wheel gear on its empennage.

25) The standard armament of the operational L-200-1 would have consisted of two 20 mm guns installed in each wing tip pod. Lockheed considered this installation the best from an aerodynamic standpoint.

NOTE 1 —
THE GUN POD DEFLECTIONS DUE TO TORSION IN THE WING ARE CALCULATED TO BE WITHIN ALLOWABLE LIMITS AT HIGH ALTITUDES AND NORMAL OPERATIONAL SPEEDS.
UNDER LOW SPEEDS AT LOW ALTITUDES AND HIGH G'S IT IS BELIEVED THAT THE TORSION IN THE WING CAN BE SATISFACTORILY COUNTERACTED BY THE USE OF THE RETRACTABLE STABILIZING FIN SHOWN IN PHANTOM. THE ADDITION OF THIS FIN CAN BE MADE, IF FOUND DESIRABLE, BY RELOCATING SOME OF THE GUN ACCESSORIES IN THE TAIL CONE. THE NECESSITY FOR AND VALUE OF THIS FIN WILL BE DETERMINED BY FLIGHT TESTS ON THE PROTOTYPE AIRPLANE.

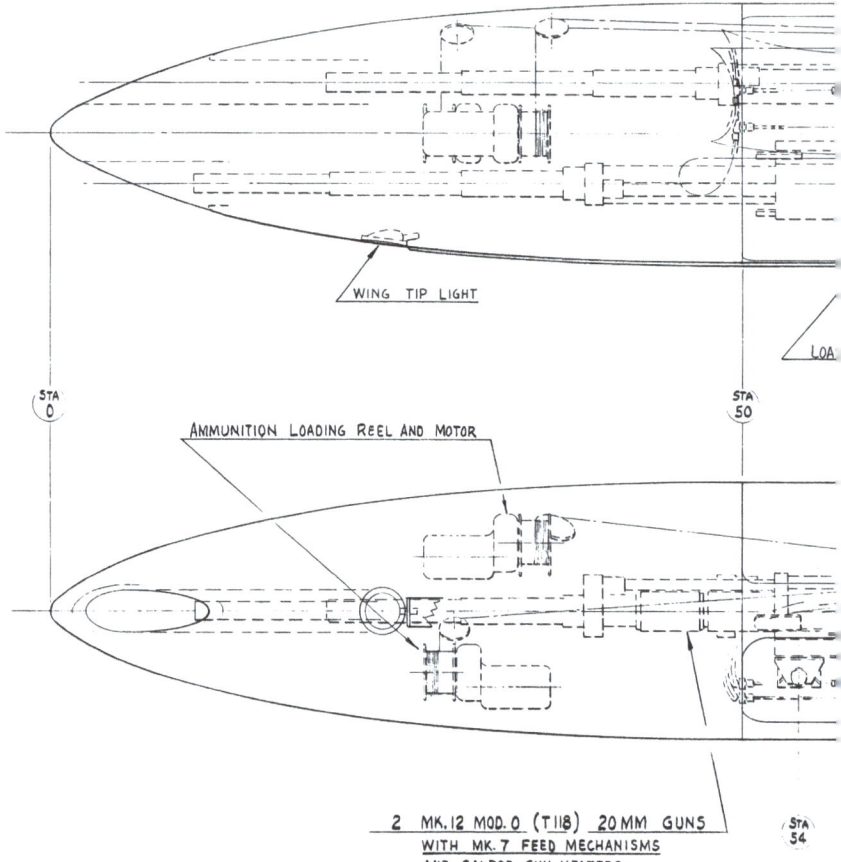

▲ 26

prototypes could be available for flight test. Under the BuAer proposal, 20 months after start of the project, the design and tunnel testing for the tactical airplanes would be undertaken, following about 2 months of flight testing. It was the Navy's desire to build the first two tactical airplanes and a static test unit on production tooling. This meant that pre-planning for production would start at the earliest possible date; in fact, almost simultaneously with the basic designing of the tactical airplane. Production tooling could not be done until tunnel tests and structural design were well along so, approximately 6 months from the start of Part II of the contract, the production tooling could start. It required at least 15 months from the start of construction of the production tooling to build the first of the two airplanes. The flight test on the tactical airplane would then start 41 months from the very beginning of the program at the very earliest.

A very considerable hazard existed in building production tooling for an airplane which had not undergone any of its experimental testing. Continued production would inevitably be faced with a flood of flight test changes. The expense involved in tooling changes would be prohibitive.

Program B presented an alternative procedure based on the conception of maintaining a given aerodynamic configuration which would gradually develop into the tactical airplane as the T-40 engine power increased to allow higher gross weights. The time required for initial tunnel tests, mock-up, and preliminary design, was the same for both Navy and Lockheed programs. Likewise, the time to flight test of the prototype was the same. The prototypes were built, how-

26) Detailed schematic of the Mk.12 Mod. 0 20 mm gun installation on the Lockheed L-200-1 Convoy Fighter.

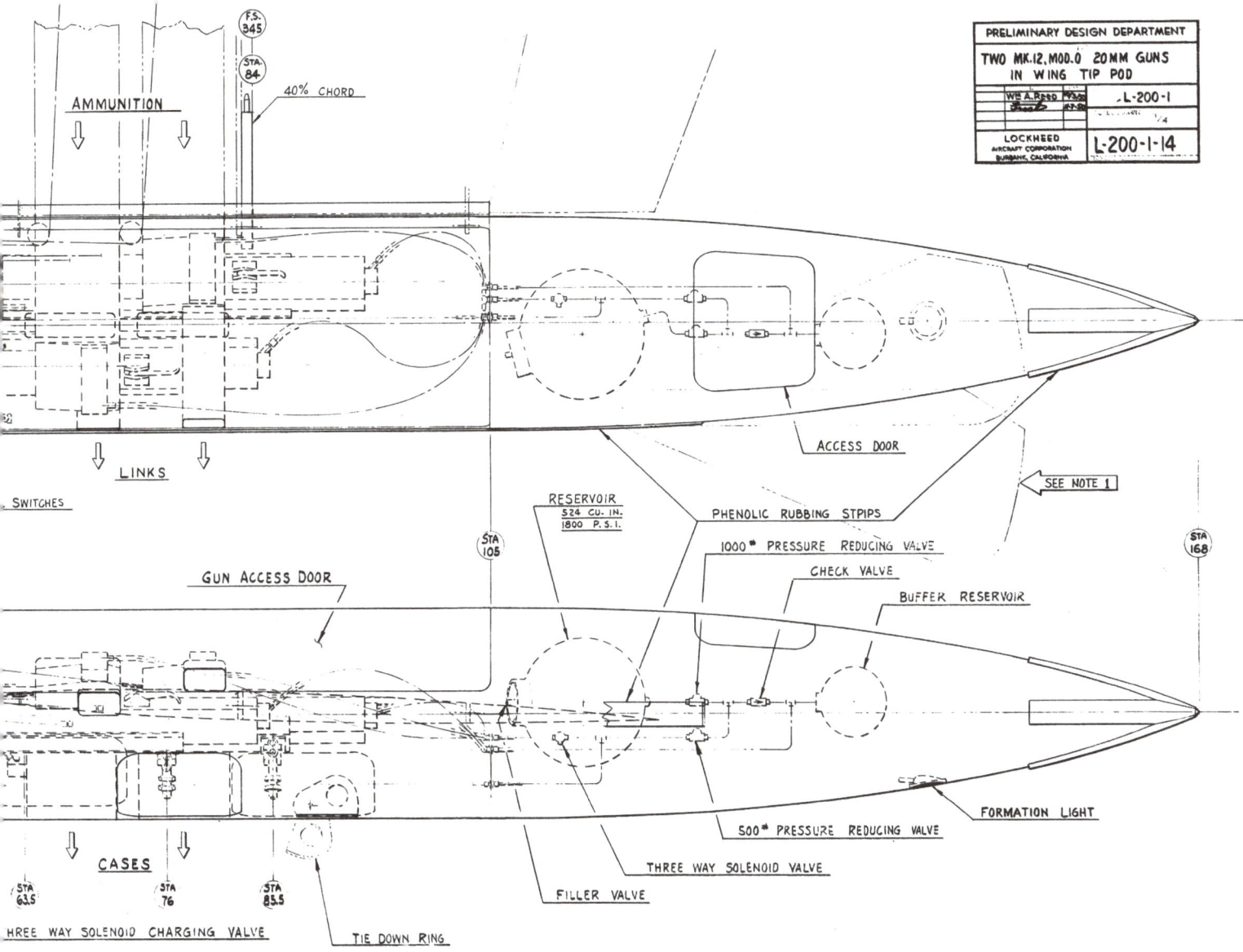

ever, on experimental tooling. This was also true for the static test article. In this program, it was advisable to have a static test article at the earliest possible date as the weight trend to the tactical airplanes was based upon obtaining the most efficient structure through continued static testing. After a period of 8 months of flight testing of the prototypes, one airplane could be converted to the tactical airplane, subject to the availability of the high-powered Allison engine. During conversion of one airplane, the other prototype would continue flight testing. It would later be converted to a tactical airplane also in a manner which would best fit in with overall conditions existing at the time. The tactical airplane could then be flown approximately 32 months from the start of the overall program compared to 41 months in the case of the plan outlined in Program A. Production pre-planning would start approximately 18 months from the beginning of the program because the basic airplane configuration for the first prototypes and the tactical airplane would have a great deal in common, particularly structure, equipment, and arrangement. Production engineering could start 22 months from the beginning of the program, or after having obtained 4 months of flight testing on the prototypes rather than 2 months as in the original plan. In spite of these advantages, it appeared that Program B would make available a tactical airplane for flight test 9 months earlier than Program A, and production airplanes would be available 9 months sooner. There would be much less risk and overall expense involved in the plan outlined in Program B than in Program A. More orderly planning and production engineering could be done with the engineering load considerably reduced by engineering only one configuration. Program B was, therefore, recommended. Cost information, however, was

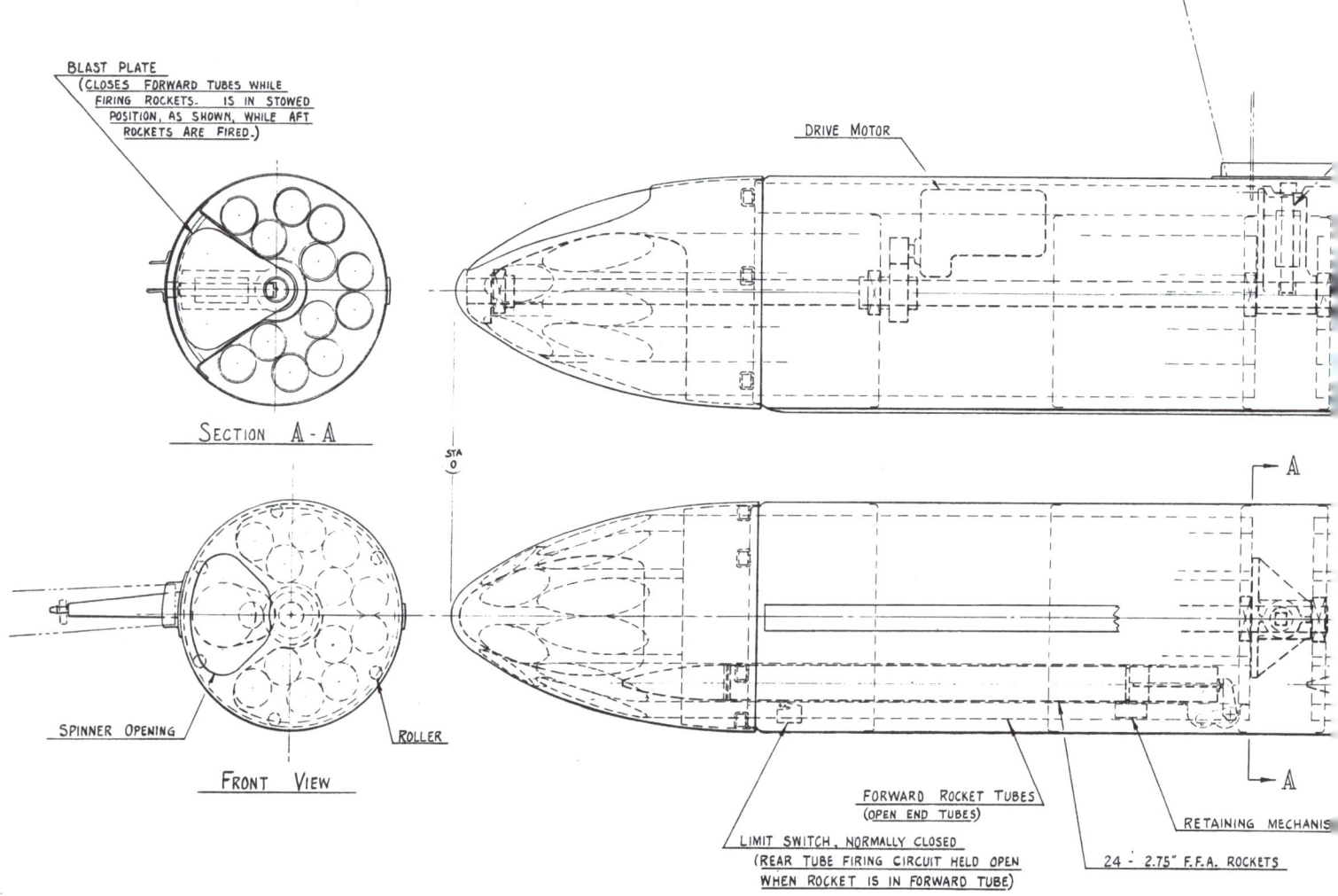

▲ 27

provided for both methods of carrying out the development.

Lockheed emphasized that it would bid and pursue whichever program the Navy considered most desirable.

Design Details

External Arrangement

The general arrangement of the L-200-1 was relatively conventional in spite of the fact that it was designed to rise and land vertically. It incorporated a straight wing, a relatively normal cockpit, external tip fairings for armament installation, and a three-way arrangement of the horizontal and vertical tail surfaces. Alternate arrangements of the airplane were studied in considerable detail and two of the most promising of these alternates are shown on pp. 16-17.

Alternate A, the L-200-6, incorporated a swept wing with 47.5° sweep of the quarter chord line and 8.1% thickness which compared to the 4% thickness of the straight wing with no other changes. The reason that this wing was not proposed as the primary configuration of the airplane was that weight studies showed the straight wing to be lighter by 210 lbs for a given drag rise Mach number than the swept wing configuration. Furthermore, since the swept wing was identical in aspect ratio to the straight wing and since its effective aspect ratio

---

27) Lockheed also studied an alternate armament of forty-eight 2.75" folding fin aerial rockets, twenty-four in each wing tip pod.

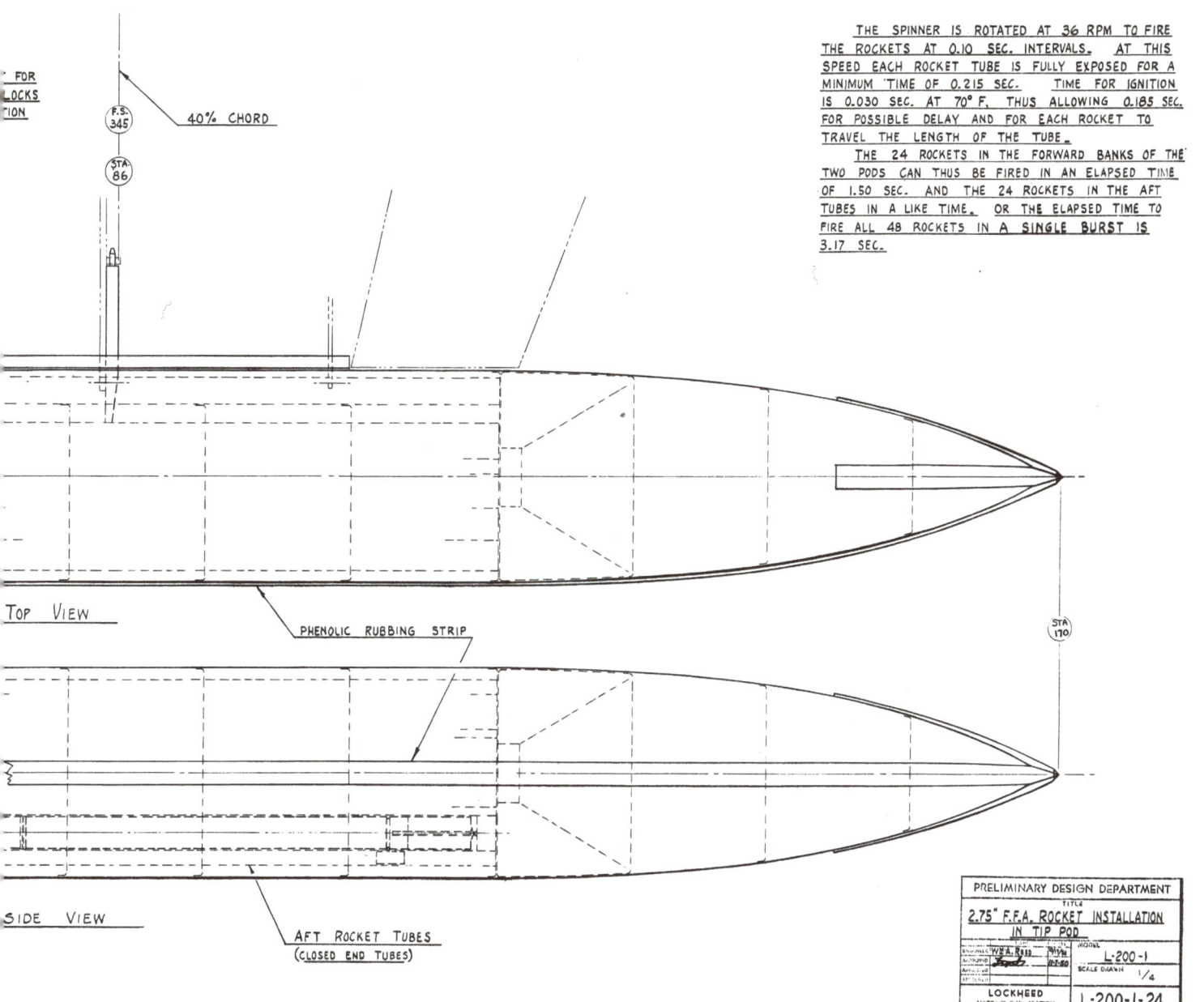

was considerably lower, the straight wing had better maneuverability at altitude and higher ceilings. Finally, the cockpit installation, control system, and armament installation were considerably simplified and the balance improved since the major structure of the straight wing went straight through the fuselage, and the gun pods were close to the airplane CG which would have been impractical with the swept wing. Lastly, it was determined that the straight wing had better aileron control in hovering than the configuration with sweep. Alternate B, the L-200-7, was an extremely attractive configuration because of its widespread footprint pattern since the wing itself was used as a landing gear support device. Lockheed recommended continued investigation of this configuration along with the swept wing configuration in the early phases of the design program because it believed that having such a widespread landing support would simplify facilities on deck and could permit simpler operation of the airplane from land bases. The major reason that Alternate B was not proposed as the primary configuration was that the stability and control, both in horizontal and vertical flight, were open to considerable question and further research was required before the feasibility of obtaining desirable flight characteristics could be proven.

Internal Arrangement

An inboard profile of the tactical airplane showing the general arrangement of equipment, power plant, cockpit and internal structure is shown at the top of pp. 20-21. An inboard profile of the proposed T-40A-6 powered prototype, illustrating the similarity between the two airplanes, is shown at the bottom of the same spread. In presenting the internal arrangement, Lockheed noted that a major revision was made to the standard gear box supplied with the Allison T-40A-8 engine in order to provide a larger offset between the power section centerline and the propeller centerline. This offset, as discussed later in the Power Plant section on p. 39, resulted in a weight penalty to the power plant section of the airplane but provided such major simplification advantages for the rest of the configuration that this weight penalty was canceled in the overall design. Some of the advantages of the offset gear box were as follows:

1. The duct inlet to the engines was shorter and had much less bend than with the standard gear box. This resulted from the fact that the spinner diameter for proper propeller efficiency was fixed regardless of the engine location and essentially covered up what would be the normal engine inlet for the standard gear box.
2. The lower position of the engine resulted in an extremely simple tail pipe installation since the exhaust was practically external on the airplane and would require practically no tail pipe.
3. Lowering the engine permitted the wing and its carry-through structure to be more nearly along the centerline of the airplane, making available a large unencumbered space above the wing for a single simple fuel tank.
4. The space available below the cockpit was considerably larger with the lower engine position, permitting mechanisms for control, seat movement, etc. more space, thereby simplifying their design.
5. With the lower engine position, the main fuselage structure was deeper, thereby effecting a major saving in fuselage weight and rigidity.
6. Having the engines partially exposed outside of the basic fuselage structure simplified and lightened the fire protection provisions which were required above and between the power sections.
7. If it was later considered feasible to add an afterburner to the airplane to improve its high speed performance, the installation of this device would be quite simple with a low engine position but would be practically impossible if the standard gear box were used.

A more complete discussion of the alternate power plant possibilities in the airplane is found in the Power Plant section. The simplicity of the service and maintenance of this airplane is well illustrated in the inboard profile, since all of the electronic gear could be arranged in essentially one place and, since the accessories as

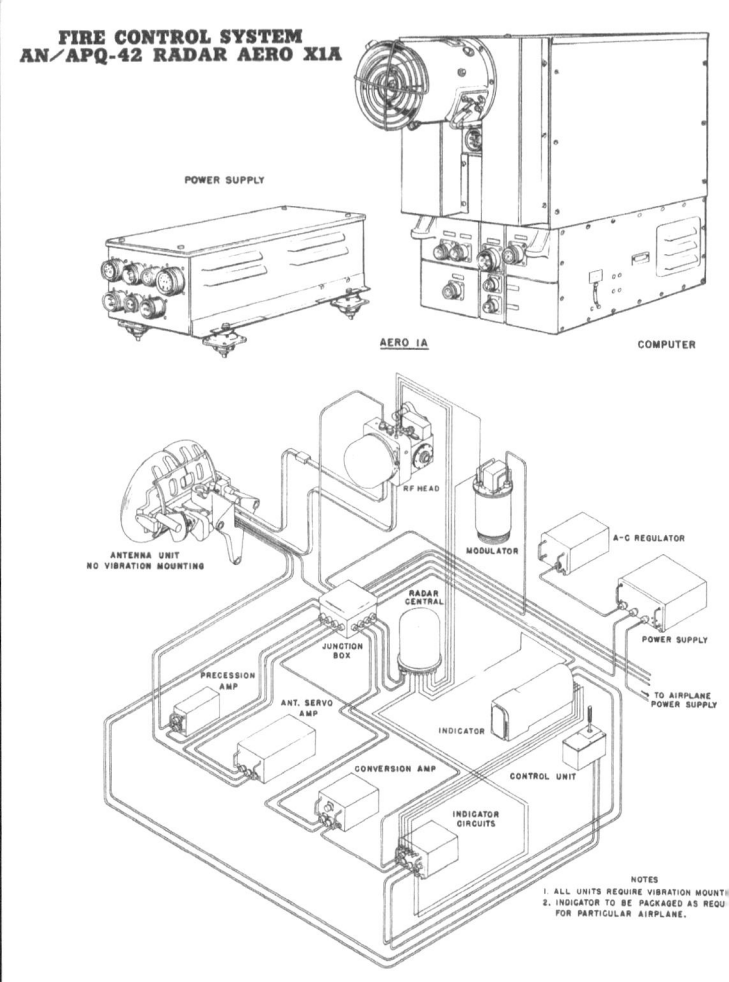

▲ 28

28) Diagram of the L-200 fire control system and AN/APQ-42 Radar Aero X1A.

29) Drawing showing the layout of the communication system antennas.

ANTENNA INSTALLATION

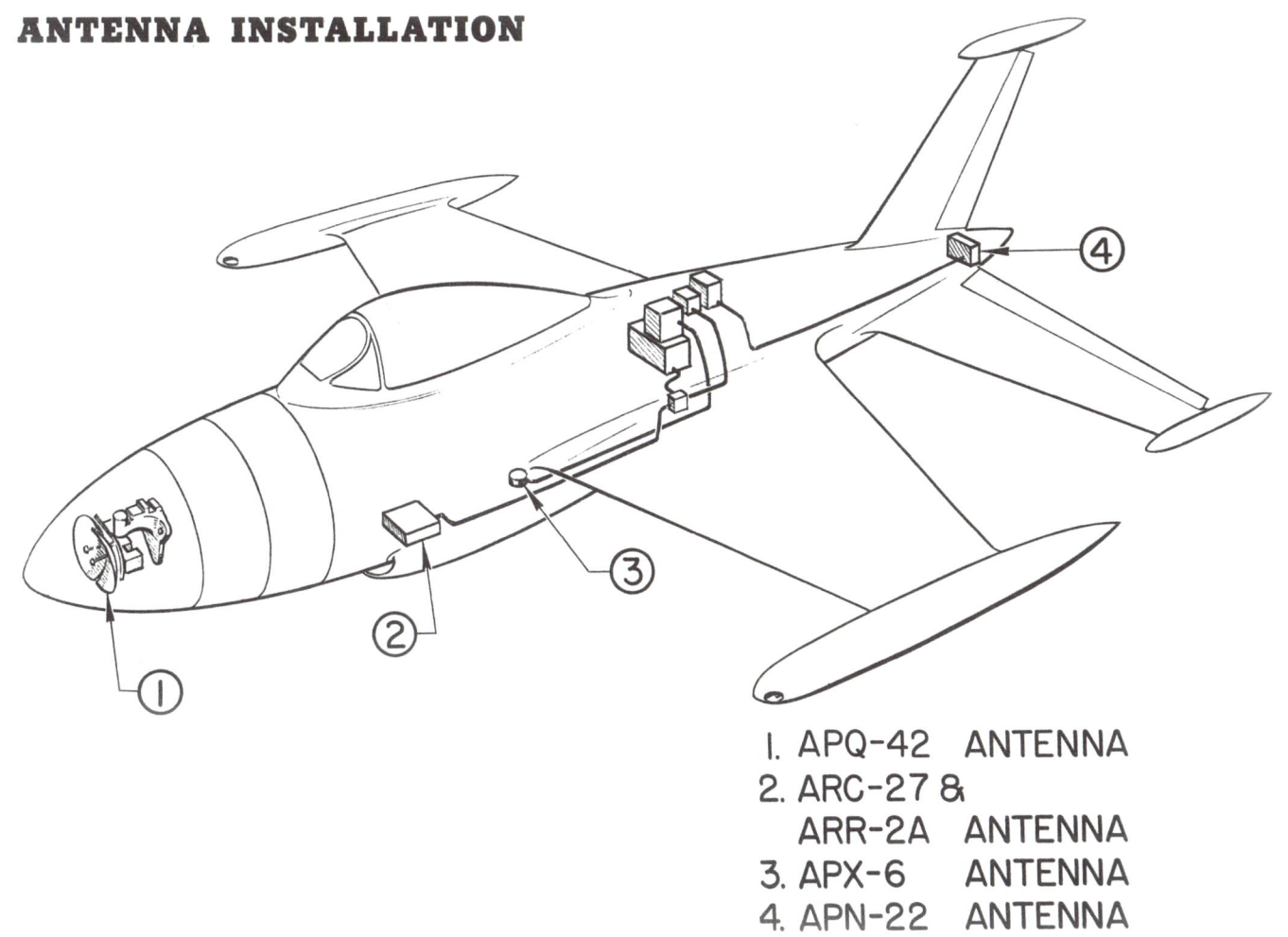

1. APQ-42 ANTENNA
2. ARC-27 & ARR-2A ANTENNA
3. APX-6 ANTENNA
4. APN-22 ANTENNA

▲ 29

proposed for the engine and the electronic gear which had to be located close to the radome, were all available through external doors without breaking through major structure or engine ducts.

Cockpit Arrangement

The design of the cockpit arrangement, as shown on pp. 22-23, followed conventional lines in spite of the unconventional requirements on the pilot. It was felt that this was very nearly the optimum approach since it corresponded to the conclusions reached by Navy investigators and since it did not impose upon the pilot unconventional positions and locations of controls and instruments simultaneously with the unconventional attitude of the airplane. Accordingly, the seat was arranged to pivot about an axis parallel to the wing axis in order to raise the pilot's head as the airplane approached its vertical attitude. This was found to be absolutely necessary from the standpoint of comfort. It was proposed that this seat position be governed entirely by the attitude gyro in the autopilot and that the movement between the normal position to the lean-forward position be gradual, starting from a nose-up attitude of approximately 45°. As a further item of comfort, stirrups were added to the control pedals in order to support the weight of the feet and legs in the vertical attitude.

Additional items of note in the cockpit arrangement included:

1. A central location for the radar scope was found between the pilot's feet with the scope face aimed directly at the pilot's normal eye position. This location did not interfere with the Mark VI Mod. 1 gunsight and it required that the flight instruments be moved only slightly to the right of the centerline of the cockpit. All of the engine and miscellaneous instruments were located on the left side of the panel.
2. Autopilot and communicating controls were located in a convenient position on the right-hand side shelves

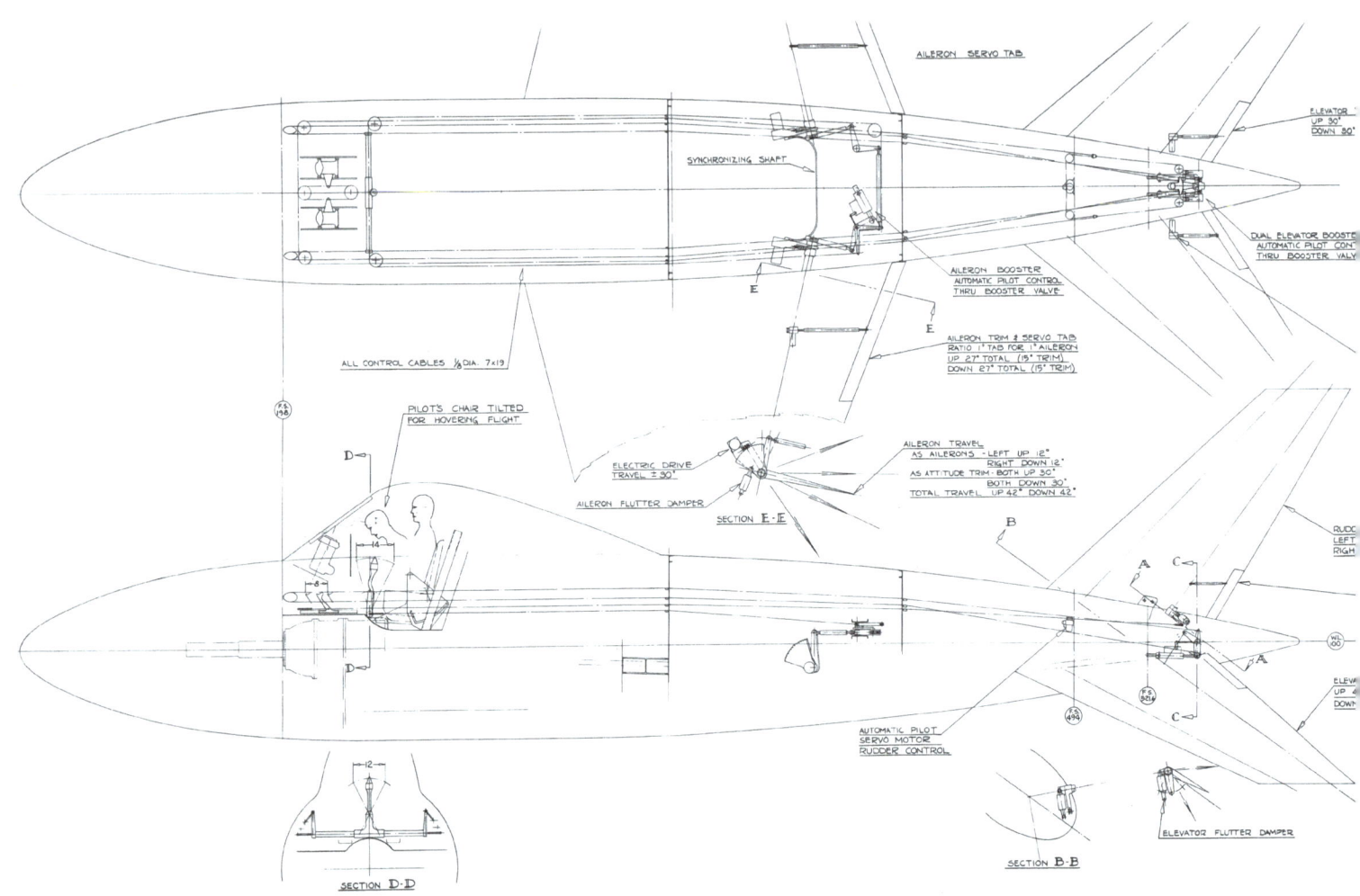

30) Diagram of the major surface control system of the Lockheed L-200 Convoy Fighter.

and were easily reached by the pilot in either the vertical or horizontal seat position.

3. The left-hand console was reserved for the power and radar controls. These also were convenient to reach with the seat in either position.
4. A separate autopilot controller for vertical flight was installed directly on top of the control stick. This control consisted of a switch which would translate the airplane in a belly-forward or aft direction by moving the thumb away from or toward the pilot, respectively. The airplane would translate laterally by moving the switch to the left or right of the pilot, depending upon the direction desired. A considerable discussion was held with several experienced pilots and the autopilot manufacturers concerning the desirable location of the autopilot controller for vertical flight and the desirable sense in which these controls should operate. This stick location was finally selected for the following reasons:
   a. Since the controller actually selected attitudes of the airplane and, therefore, as a result, selected translational velocities. It was believed that a sidewise motion of the controller was more suitable for side movement of the airplane than the yawing type of control normally associated with an autopilot controller which turned the airplane to the left or right.
   b. Thumb movement away from the pilot for forward motion and back toward the pilot for backward motion was believed to be in the satisfactory sense since this corresponded to the motion of the stick to achieve fore and aft translation.

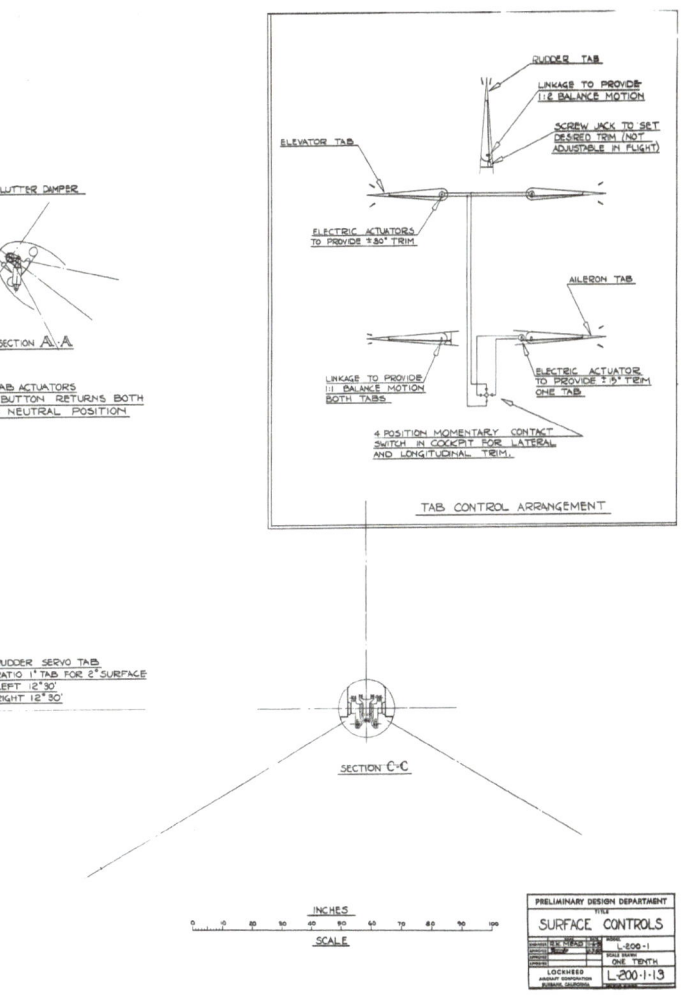

    c. Although similar motions could have been worked out in a very reasonable fashion on the standard autopilot controller by changing the control head, it was believed to be impractical to remove the pilot's hands from the control stick in order to achieve this simplification since a failure of the autopilot might occur at such a time that the pilot could not afford the delay of moving his hands from a separate autopilot controller to the stick in order to retain control of the airplane.

5. A radio altimeter was placed in the lower left-hand corner of the instrument panel with its face attitude adjusted for easy reading by the pilot when the seat was in the vertical position. It was been found that this location of the altimeter was the most desirable when the pilot was flying the airplane in its vertical attitude since it required the minimum change in eye position from the normal outside vision line past the wing tip toward the ship

Other seat positions, including 180° rotation positions, cocked axis positions, etc. were all investigated in an attempt to improve the comfort and usability of the cockpit. None of these seat ideas achieved any advantage except that of better and more comfortable direct rearward vision. It was concluded after these mock-up studies that this vision, when obtained, was not nearly as valuable as maintaining a normal cockpit with normal controls since the accurate vertical height adjustment of the airplane could better be handled through the inclusion of an LSO on board ship who could better judge the actual motion of the ship and more accurately determine the precise altitude for power cut. For operation from land bases where no ship motion need be accounted for, Lockheed believed that the pilot could adjust himself sidewise in the existing seat to obtain sufficient down-vision. Furthermore, if the landing platform as proposed for the ship was utilized for ground bases, an altitude target directly to the pilot's left and just beyond his wing tip was provided by the wing tip backstop.

It was believed that the pilot would have a more accurate determination of altitude at touchdown from such a target than he would from direct aft vision even under the most ideal location of his seat.

Power Plant

A general arrangement drawing of the final power plant as proposed for the L-200-1 airplane is shown on pp. 24-25. The basic power sections were those of the Allison T-40A-8 engine as described in Allison Specification No. 272B, revised 31 May 1950. The power ratings of the engine used for the performance of the airplane were:

Military Power	6,955 shp at 14,300 rpm
Military Thrust	1,363 lbs
Normal Rated hp	5,790 shp at 14,000 rpm
Normal Rated Thrust	1,225 lbs

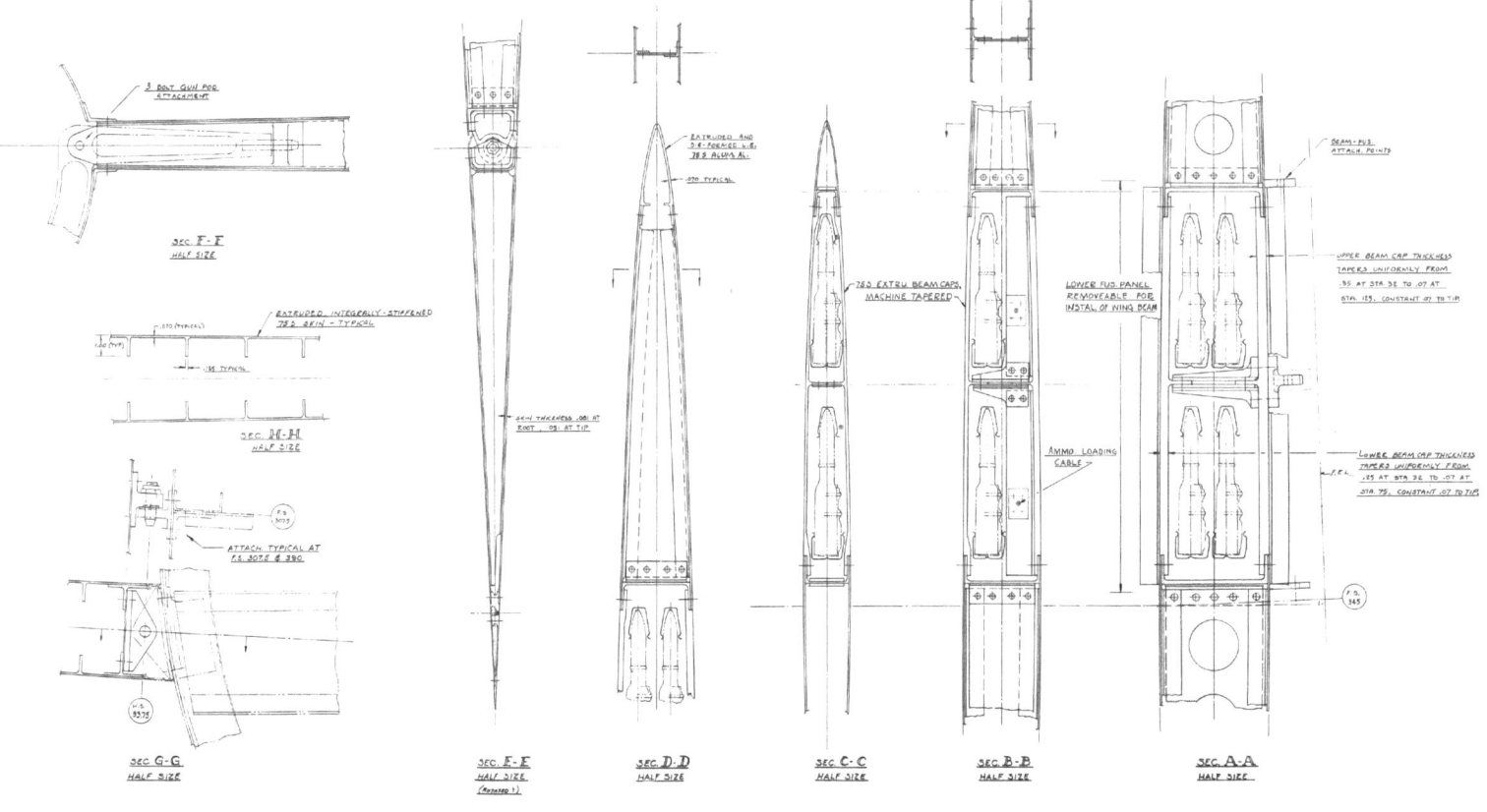

This engine was modified by changing the standard single speed gear box to a two-speed gear box with a larger offset between the power section centerlines and the propeller centerline. A discussion of the reasons for this additional offset was given under the "Internal Arrangement" section on p. 35. Since this gear box was the subject of considerable discussion between BuAer, the Lockheed Company, and the Allison Company, a clarification of the gear box development program was obtained from Allison. As a result of this investigation, it was decided to use Gear Box No. 3 shown in the table on p. 29 since its development time did not exceed that of the standard gear box with the two-speed unit and since it had a sufficient offset to improve the airplane arrangement in a major fashion.

Study of the power plant and aircraft accessories indicated that the most desirable accessory location would be on the front side of the gear box where all of the accessories for the airplane would be reached through external doors or through the duct closure doors. Since this location of accessories caused an additional development period on the gear box, it was decided to forego these advantages in the proposal and accept Gear Box No. 3 with the normal accessory positions on the aft side. Further study of this item was required during Phase I of any contract to determine whether the forward location of the accessories actually necessitated the additional gear box development time or whether the program of aircraft development could reasonably stand such a delay.

During the analysis of the airplane performance, a further discussion was held with the Allison Company and BuAer as to the feasibility and desirability of using an afterburner to eliminate the necessity of a two-speed gear box in the airplane, as shown on pp. 26-27, or in addition to the use of the two-speed gear box. It appeared that the use of an afterburner would improve the high speed performance by approximately 13 kts without the two-speed gear box and 38 kts if both the two-speed gear box and afterburner were installed. These high speed advantages could only be obtained, however, with a weight empty penalty over the proposed airplane of approximately 435 lbs with the single-speed gear box or 635 pounds with the two-speed box. This included a weight penalty of approximately 498

31) **This structural diagram of the L-200-1 wing shows the major improvement in simplicity and structural integrity enabled by the elimination of the landing gear, armament, fuel, and flap provisions within the wing.**

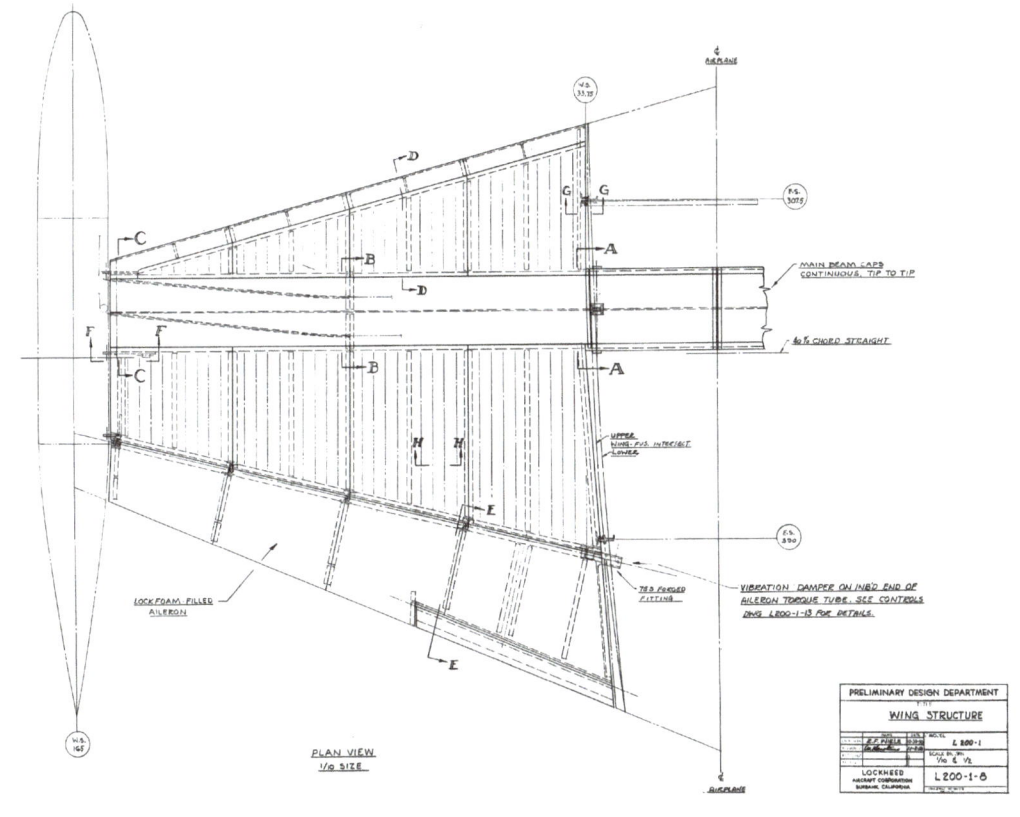

1. A single power section failure or cutoff for cruise purposes.
2. Plenum chamber evacuation due to suction from the power plants at zero forward velocity.

A final deterrent to the use of this type of duct was the inability to separate the two power sections to prevent backfires from causing a fire hazard in a plenum chamber containing accessories and inflammable materials.

Lockheed believed it desirable to have two power sections in a vertical rising airplane and, therefore, the T-40 type power plant arrangement was preferred. Since the development of propeller turbine power plants, however, had been extremely complicated and difficult, it was believed to be important that an alternate engine be available to do the job since this airplane could depend neither on an alternate reciprocating engine nor on an alternate pure jet engine.

Lockheed provided a summary of power estimates with time for the Allison T-40 engine, showing the availability of the power ratings used in Lockheed's proposal from a date standpoint. Plotted also on this curve was a power development proposal for the Pratt and Whitney PT2E engine. This indicated the feasibility of incorporating a Pratt and Whitney engine in an airplane of this type at approximately the same time. The airplane could be made entirely usable and have similar performance with the Pratt and Whitney engine. This required, of course, additional gear box development to obtain a two-speed offset counter-rotating gear box but such a development would probably not be as difficult as the similar development incorporating the two power sections.

lbs of fuel which was required if the afterburner was used for climb and combat. Only 282 lbs of this fuel could be conveniently carried. Lockheed recommended that this type of power plant also be considered in any Phase I study in order to determine whether or not the high speed advantage was worth this penalty and whether or not it was necessary to incorporate two speeds in the propeller gear box where an afterburner was installed.

As shown on pp. 24-25, the ducts for the oil coolers and the ducts for the engine intake were separated in order to ensure that flow through the oil coolers would always occur in one direction and would not be disturbed by variation in flow into the power sections. Several alternate schemes were studied in which the oil coolers were fed by a large plenum which also fed the inlet to the engines. This system had several advantages since it eliminated the fuselage cutouts for the oil coolers, permitted the oil coolers to be located further forward in the airplane and more equally distributed the air entering the engines. It was discarded, however, since a major thrust loss was calculated for this type of plenum chamber and it was impossible to determine what kind of pressure ratio would be available for oil cooling in the event of:

Airplane Landing Provisions

As described previously under the "Operational Features" section, the landing with this airplane was made by flying sidewards into a shock absorbing wing tip stop under the direction of an LSO and, upon being given the cut, the power was reduced and the airplane dropped into a landing net on the three-pronged empennage with the spike from one of the wing tip pods penetrating a small wing tip net to provide stability against inadvertent tipping over of the airplane due to ship roll or excess side velocity at touchdown. The provisions for landing on the tip pod consisted only of a rub strip of either hard wood or titanium attached to the outside of the normal tip pod and rub ribs near the tail of the pod to prevent scraping by the tip net cables. Landing provisions on the tips of the vertical and horizontal tail, however, were more exacting and required not only pads to prevent the tail from penetrating the landing net but also required shock absorption means to prevent damage to the airplane and reduce the load factor which needed to be carried by the empennage.

The typical tail tip pod had a mechanism for retracting the landing pads and obtaining adequate shock absorption. The facilities for landing the airplane, though heavy, were not excessive when compared to a normal landing gear for an airplane of this weight. Alternate landing procedures for prototype or land-based flying were studied, and Lockheed believed that none of these procedures could easily be incorporated into the airplane as a permanent type of landing system, since they involved complication and weight which would seriously impede its tactical utility. Lockheed developed alternate wheeled gear to show what could be done with the airplane during the early development stages of flying. Lockheed designed a very simple gear for operating the airplane in a normal fashion without reverting to vertical flight for landing or take-off. The gear shown on p. 30 is another type of interim installation which could be used for

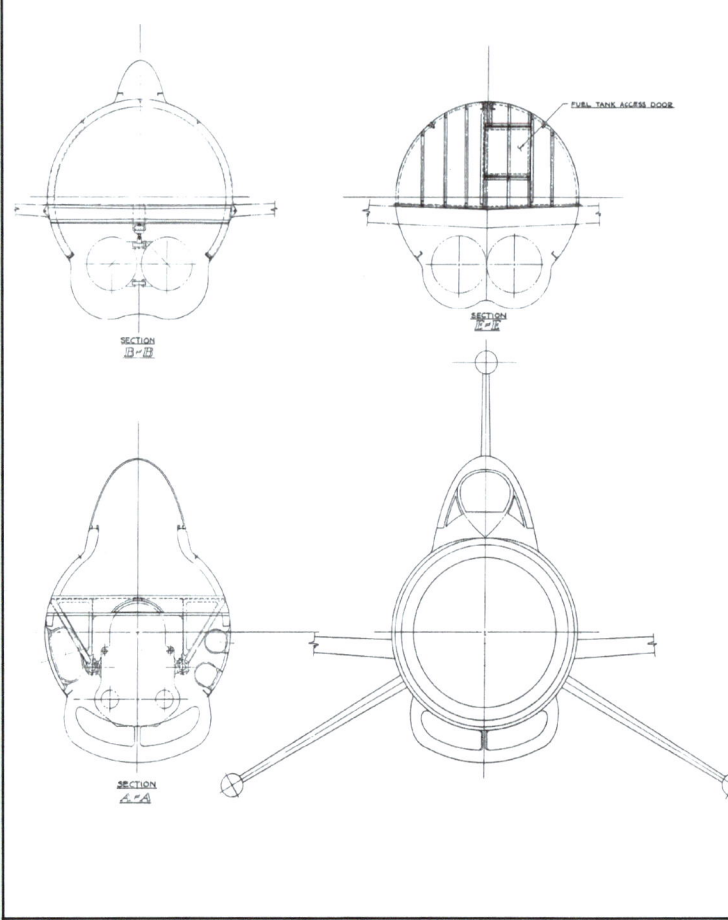

▲ 32

the first vertical flights of the airplane or prototype while developing the simpler tactical system.

Armament

The installation of the four 20 mm guns required in OS-122 appeared to be optimum when a wing tip pod was installed on the airplane, as shown in the drawings on pp. 31-33. Installing this pod on the extreme tip of the wing was obviously best from an aerodynamic standpoint and it had many additional advantages when compared to any other arrangement for the guns. These advantages included the following:

1. Permanent location of the guns at the extreme tip of the wing saved approximately 440 lbs in the wing bending material.
2. The gun location in these pods made access for removal and loading as nearly perfect as could be obtained.
3. By locating the guns at the extremi-

32) This structural diagram of the L-200-1 fuselage shows that a major simplification was achieved due to the airplane's VTOL tailsitter configuration and the incorporation of a gear box with a large offset.

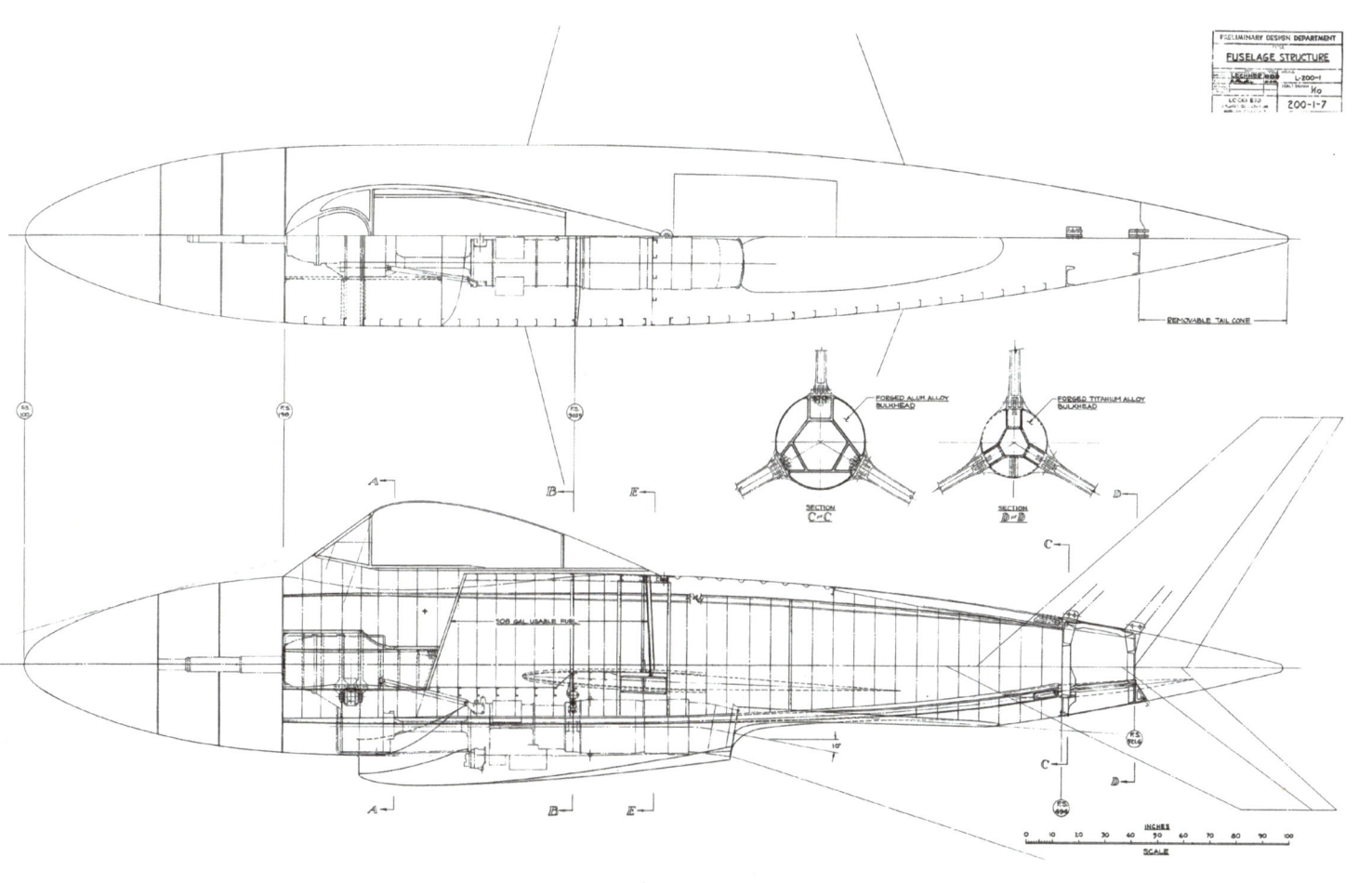

ties of the wing, the entire wing span became available for simple integral ammunition chutes.

4. The existence of the wing tip pod provided an external stabilizer prong for the landing system and an external buffer for the wing tip backstop.
5. When combining the wing tip installation with a straight wing, the CG advantage of having the ammunition practically on the center of gravity of the airplane was very apparent.
6. Locating the guns and pods near the airplane center of gravity made it simple to change the type of aircraft armament at will without seriously affecting the airplane design. As an instance of this, the blueprint on pp. 34-35 was included, showing the ease with which 48 folding fin rockets could be installed in lieu of the four 20 mm guns. This installation affected none of the basic airplane design and, in fact, it appeared feasible to have alternate wing tip pods for installation on the same airplane.

Loading of the ammunition for the tip guns was done by means of a built-in electrically powered ammunition loading cable. This cable was attached to the end of the belt and the belt was pulled into the wing by the activation of the control button on the tip through which the ammunition was being led. Since the electric winch for this ammunition loading was installed in the opposite tip pod, no difficulty was encountered if it had to be replaced. Furthermore, the cable, if frayed, could easily be replaced by threading through the straight ammunition chutes. The cable remained attached to the ammunition belt and was immediately available at the wing tip for reloading when an ammunition belt had been expended.

One problem introduced by the tip pods was the ability to maintain boresight accuracy during maneuvers at high speeds and high load factors. For extreme altitude operation where load factors were low and the Mach number was high, Lockheed found that tip deflections were

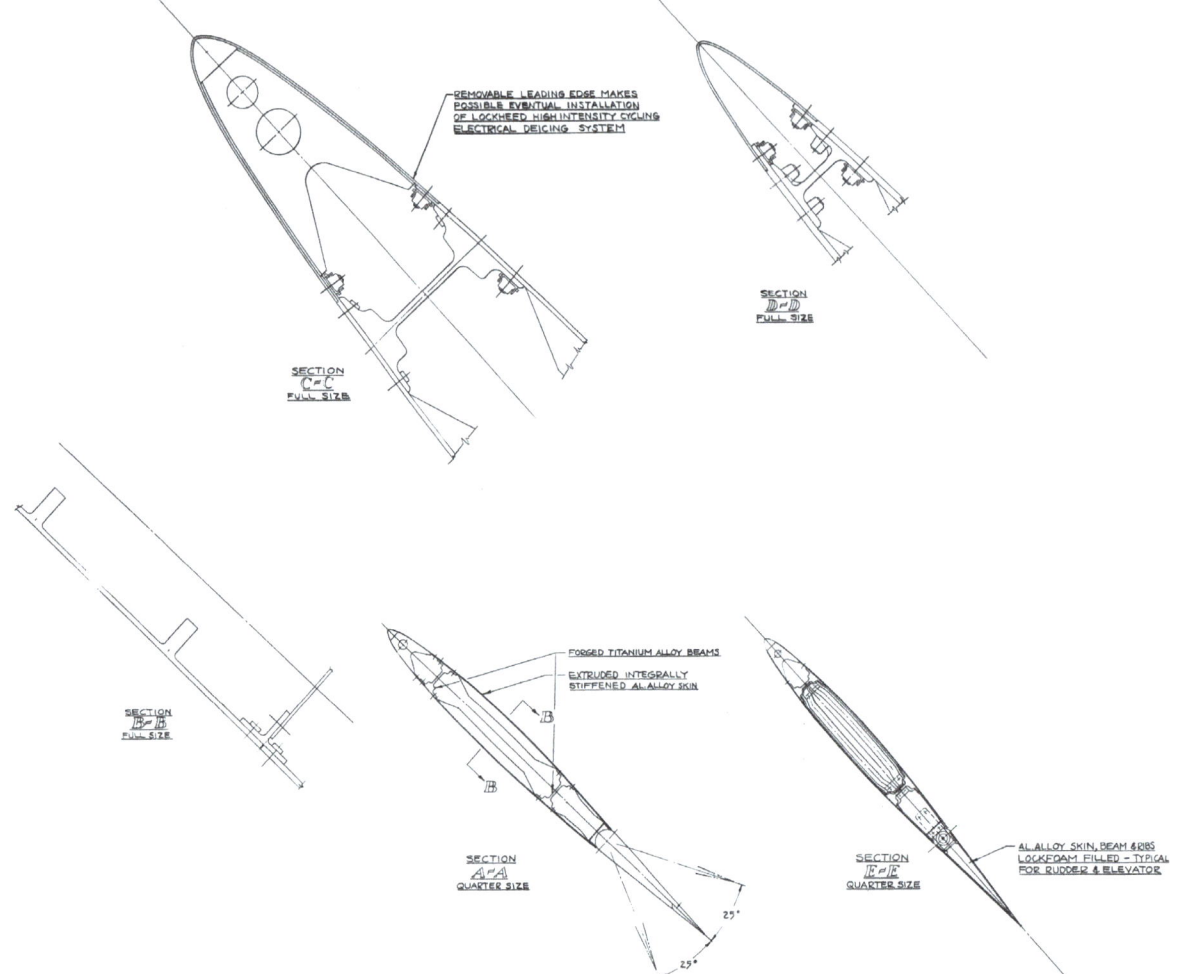

▲ 33

within the desired accuracy. However, low speed high load factor maneuvers at sea level were believed to cause some divergence. A solution to this problem was suggested by contemporary high load factor maneuvers with tip tanks equipped with tail fins. These indicated that tip torsional deflections could essentially be eliminated and, as a result, a tail fin option was proposed by Lockheed.

Electronic and Electrical Provisions

The autopilot installation for the Lockheed L-200 was probably the most important item of electrical equipment since its proper functioning and installation ensured the ease of making the flight transitions in an ideal fashion and in effecting the landing and takeoff satisfactorily. Discussions were held with several autopilot manufacturers and it appeared, for proposal purposes, that the Westinghouse W-3 autopilot, suitably modified, met all of the requirements. The major component of the autopilot was a rate gyro system which was used under normal flight operations. The added feature for use in vertical flight was the vertical gyro. As described previously, this gyro was used to ascertain the vertical axis about which the airplane was maneuvered when in a vertical position. A signal from the autopilot controller on the stick asked for a zero rate of motion about a fixed airplane attitude in space, either laterally or in a pitch direction. This attitude was held through the rate gyro system whose signals were implemented by a teeter valve on the elevator boost control and by servos on the rudder. Zero roll rate was maintained for all attitudes through an aileron servo. The system which eliminated a servo on the elevator control had been checked out on flight tests of the Lockheed F-94 all-weather fighter airplane, and performed in an entirely satisfactory man-

33) The L-200 empennage presented the greatest structural difficulty with the VTOL tailsitter configuration; not only did the tail have to withstand the air loads but also the landing loads. It had to be both strong and light; to achieve this, Lockheed proposed utilizing titanium forged empennage spars and an aft fuselage attachment bulkhead.

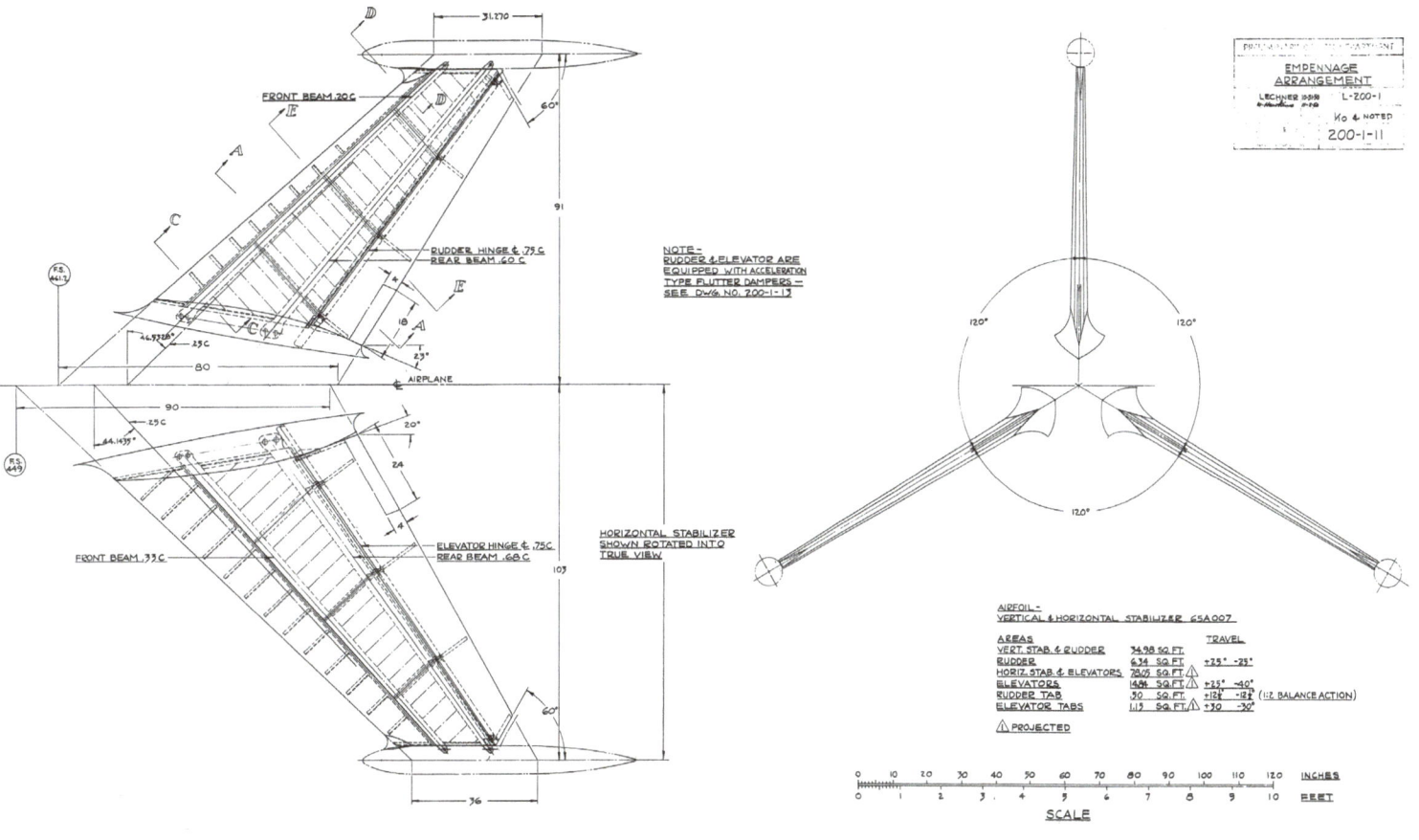

ner.

Aside from having all of the components which were required for vertical flight, this autopilot also provided a controller which could be used by the pilot in gunlaying and tracking maneuvers for combat purposes. Under these circumstances, the normal autopilot controller on the right-hand console was used.

The figures on pp. 36-37 illustrate the remaining electronic components of the airplane, showing the relative locations of the radar scanner and the various components which had to be located throughout the L-200-1 in order that the radar may be satisfactorily operated. As noted previously in the "Internal Arrangement" section, all of the components of the radar were available for simple servicing and maintenance procedures. The power system for operating the electrical and electronic equipment had as its main power source a 15 kva alternator driven from the standard alternator drive pad on the rear of the propeller gear box. Other alternate electrical systems were studied but it appeared that this was the lightest and most usable, considering the limitations on the power drives which could be available with reasonable development time.

The diagram on p. 37 of the communication system antennas shows the following required communication and navigation sets:

AN/ARC-27	UHF Command Set
AN/ARR-2A	Navigation Receiver
AN/APX-6	IFF Transponder

These sets were those listed in OS-122, and the ARC-27 essentially limited the operating range of the airplane to approximately 120 nautical miles at the operating altitude of 35,000 feet. As discussed under "Operational Procedures," it appeared that the communication range of the airplane was the actual limiting factor if the airplane was used for search purposes. Therefore, a Phase I study of the actual operational use of the airplane required a detailed study of its utility as a search airplane in order to determine whether or not recommended changes in communication equipment were desirable. An alternate gunlay-

45

ing and radar system had also been studied. The utility of this radar and gunlaying system was that it permitted the airplane a larger flexibility in the final attack maneuvers since it permitted collision course flying as well as tail chase operations. With this equipment it was possible to carry and effectively use 2.75" folding fin rockets. Lockheed believed that this also required study in considerable detail during the early phases of the contract development of the Convoy Fighter since the margin in time by which a 450 knot bomber was intercepted was extremely small and collision course attacks would materially improve the pilot attack procedures and shorten the attack time.

Control System

The diagram on pp. 38-39 shows the major surface control system of the L-200. The control system featured the following basic ideas:

1. The elevator control system comprised two completely independent power boost units so arranged that battle damage could not reasonably eliminate both units. This eliminated the necessity of a complicated and undesirable aerodynamic balance system which would have to be devised for boost-off flight and which could not reasonably be made to control the airplane over its entire speed range.
2. The aileron control system utilized a single boost cylinder for normal flight operation and incorporated a servo tab located in the propeller slip stream for boost-off operation. This servo was not completely satisfactory for combat flying and, therefore, the boost could not be eliminated but it did serve to give adequate aileron control during the flight transition between horizontal and vertical position and did give adequate aileron control during vertical flight. An aile-

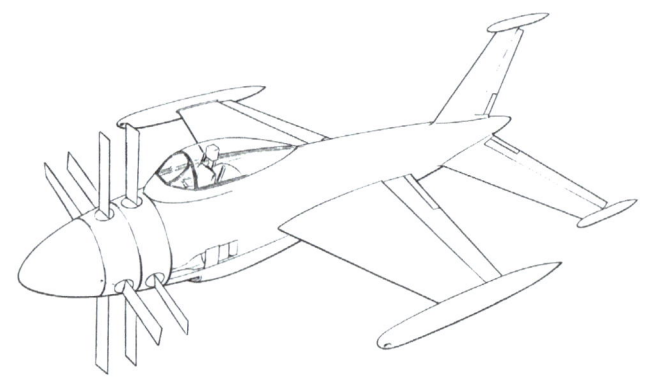

▲ 34

ron droop or flap movement had been incorporated to act as a vertical flight attitude trimmer. It was found from wind tunnel tests that cross winds perpendicular to the wing plane required high tilt angles to position the airplane. A full-span flap alleviated this condition and permitted the tilt of the airplane to be approximately the same in pitch as in yaw for a given translational velocity.
3. The rudder system incorporated a servo tab and required no boost since satisfactory pedal forces could be achieved for both combat and transition flying.

The trim control system was conventional in every respect. It was notable that the vertical flight operation of the airplane eliminated

---

34) The L-200 production breakdown was identical for both the full-size prototype and production airplanes, and was based on service, spare part requirements and producibility.

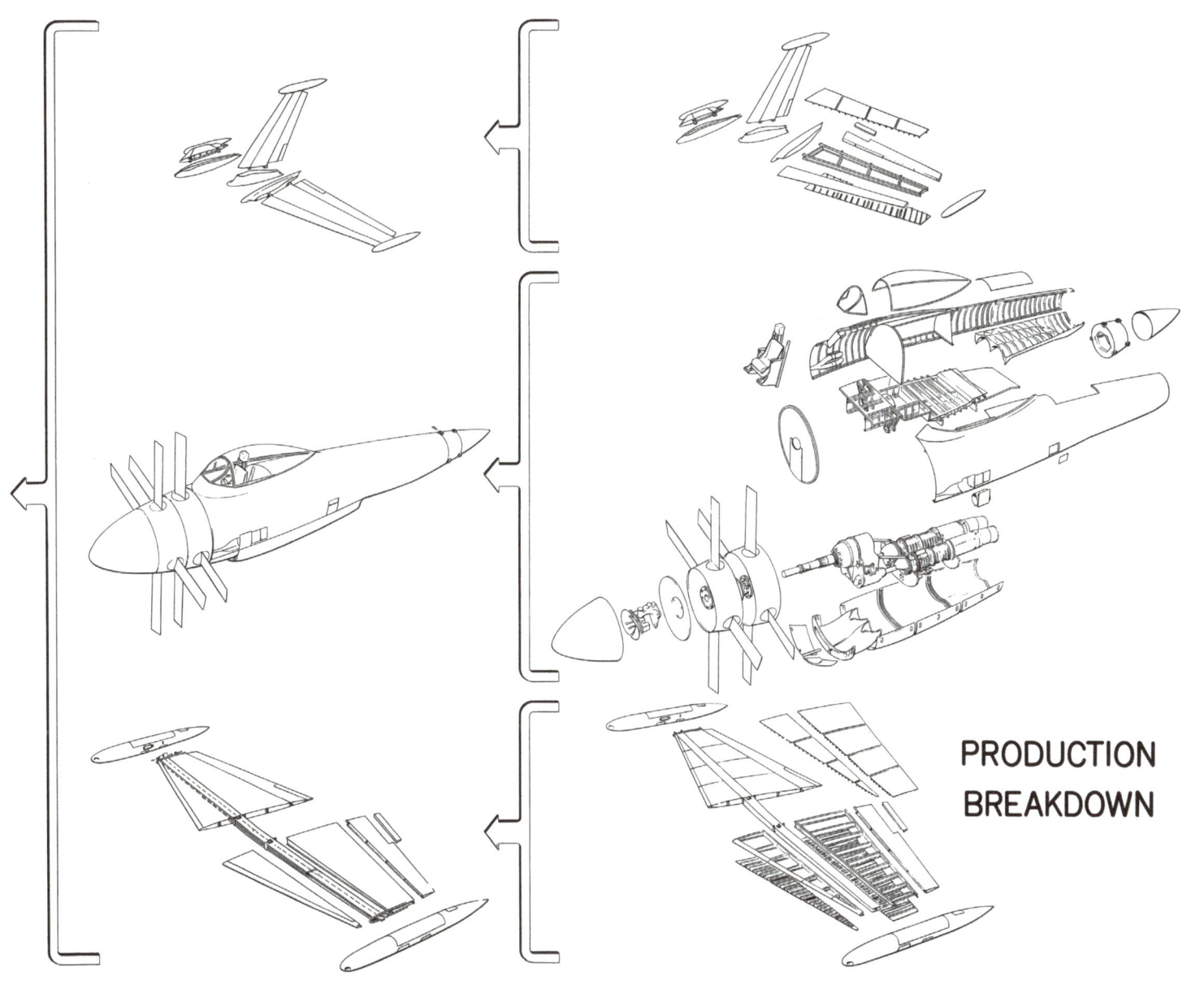

PRODUCTION BREAKDOWN

some of the most severe trim conditions since no major moment change existed which was comparable to the operation of the landing flap for a normal airplane. Lockheed emphasized that the lowered position of the engine made possible by the special gear box simplified moving the controls out of the cockpit and arranging them within the airplane. This resulted in major weight savings and major improvements in simplicity which were directly responsible for easy servicing and infrequent maintenance attention.

Studies of the flutter characteristics of the control surfaces indicated the need of closely controlled dynamic and static balances since the flight speed range of this airplane caused the flutter problem to be relatively severe. The solution of this problem in the normal fashion by the addition of weights was hampered by the thinness of the control surfaces and the desire to maintain the maximum possible control requiring sealed hinge surfaces. This led to the incorporation in the control system of dynamic dampers in place of the normal weights mounting ahead of the hinge line. The dynamic damper was developed by Lockheed in its study of the phenomena of wing aileron buzz and was a simple self-contained unit which permitted normal rates of movement of the control surface but did not permit rates of motion of the surface above those amounts. An actual flutter damper was put through many exhaustive tests in the Lockheed Research Laboratory, including tests at low temperature, at high temperature, and for prolonged periods of operation. The basic principle of the unit involved a frictionless lead screw (ball bearing race) driving a rotating shaft. On this shaft was a rotating collar driven through another frictionless lead screw. With this arrangement

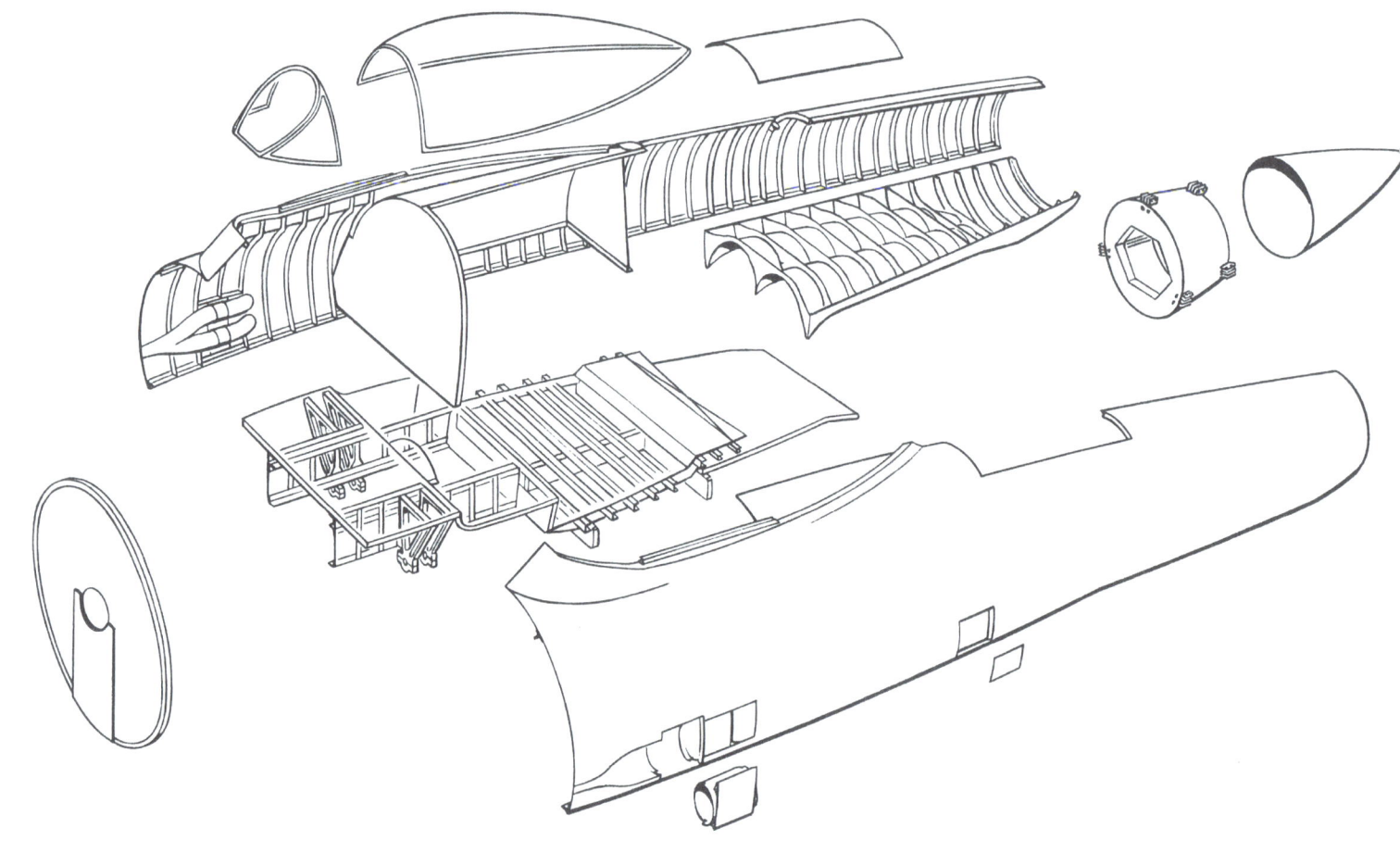

FUSELAGE BREAKDOWN

▲ 35

the collar rotated at the same speed as the shaft if no acceleration or deceleration took place. If acceleration was present, the shaft rotated faster than the collar and the outside lead screw then drove the collar against a thrust bearing which applied a brake to the entire system. Since the brake had a double surface, the reverse took place during deceleration. This braking system was found to be entirely effective in eliminating buzz and flutter. An alternate scheme to solve this problem would have been the incorporation of irreversible boost components in which two boosters per control surface were installed, either one of which would be satisfactory for flutter control and surface control in the event of boost failure. Since this system required the installation of false feel or the elimination of any pilot feel, it was felt that it was the least desirable of the two because anything which was uncon-

35) The L-200 fuselage broke down into five major assemblies: left and right side panels, floor and engine mount sections, lower aft section, and empennage attaching section.

ventional with regard to flying this airplane should be eliminated until the ease with which vertical flight could be accomplished was determined.

Hydraulic System

As noted previously in the "Control System" section, hydraulic boost was installed on the aileron and elevator and a dual boost was installed on the elevator. This necessitated two completely independent sources of pressure. This was achieved by driving the main utility system from the one hydraulic pad which was available on the engine and driving the other system through an air driven hydraulic pump connected with air bleed holes on both power sections of the Allison engine. This power system was made as simple as possible to eliminate any possibility of failure since it was the relief system upon which the elevator control depended on the event of main system failure. As pointed out previously, no connection whatsoever existed between the two hydraulic systems, thus ensuring that no failure would occur which would eliminate elevator control. Lockheed recommended an engine driven pump for the reserve boost system if additional pads became available on the gear box.

The basic system used 3,000 psi hydraulic pressure and was of the constant pressure type. The lack of landing gear and flap operating devices made the system extremely simple and no unconventional features existed which could cause service trouble.

Air Conditioning System

No new components were required for the performance characteristics and altitudes expected of this airplane and, therefore, no development program was envisioned other than a test checkout of the air conditioning system and its mechanical components. Power sources from both power sections were utilized with a crossover valve which permitted cabin pressure and anti-fogging even though one power section was inoperative. The cabin pressurization was set for a differential pressure of 3.3 lbs per square inch under standard operating conditions, and 1.3 lbs per square inch for combat conditions as per Navy Specification SR-163a.

Basic Structure

Wing Design

A structural diagram of the L-200-1 wing is shown on pp. 40-41. This structure graphically illustrates the major improvement in simplicity and structural integrity which was made possible by the elimination of any landing gear, armament, fuel, and flap provisions within the wing. Further augmenting this simplicity was the decision not to use sweep to obtain low drag at high speeds but rather to attempt the use of extreme

36) Table showing the breakdown of all the required weights for performance calculations based on the recommendations and definitions of OS-122.

37) A complete summary of the L-200 weight empty breakdown estimate based on the suggested prototype program.

GROSS WEIGHT SUMMARY - L-200-1

WEIGHT EMPTY		11,315
Pilot	200	
Oil	69	
System Fuel and Oil	57	
(4) 20 mm Guns	450	
600 rds. Ammunition	408	
Oxygen and Equipment	53	
EQUIPPED WEIGHT		12,552
40% Fuel (203 gals.)		1,218
LANDING WEIGHT		13,770
20% Fuel (102 gals.)		612
COMBAT WEIGHT (Design Gross)		14,382
40% Fuel (203 gals.)		1,218
TAKE-OFF WEIGHT (508 gals. fuel)		15,600

▲ 36 ▼ 37

WEIGHT EMPTY SUMMARY - L-200-1

WEIGHT EMPTY		11,315
Wing Group		967
Panels	836	
Ailerons	131	
Tail Group		792
Stabilizers	498	
Elevators	65	
Fin	191	
Rudder	38	
Fuselage		1,250
Alighting Provisions		300
Power Plant		5,777
Engine (XT40-A8)	3,310	
Accessories	34	
Power Plant Controls	37	
Propellers	1,918	
Starting System	44	
Lubricating System	177	
Fuel System	257	
Fixed Equipment		2,229
Instruments	128	
Surface Controls	311	
Hydraulic System	163	
Electrical System	303	
Communicating	216	
Armament Provisions	803	
Furnishings	278	
Anti-Icing	27	

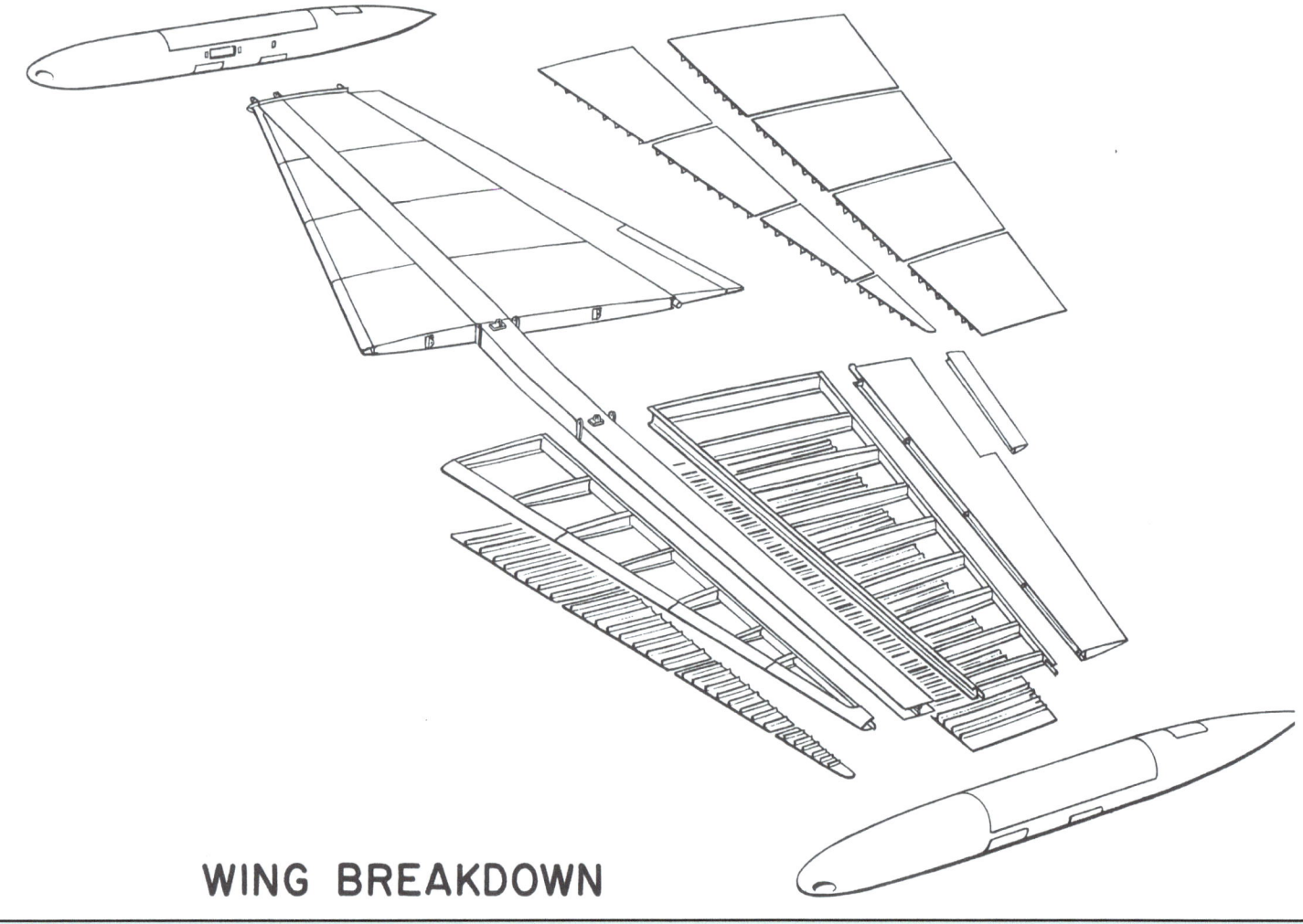

WING BREAKDOWN

▲ 38

thinness. This effort was amply awarded with the straightforward structure which was light and simple to produce. Since the problem of wing design was primarily structural and structural alone, many alternate possibilities suggested themselves to improve weight or to improve cost of construction. As previously mentioned, the location of the wing tip gun pods was extremely important since it eliminated approximately 440 lbs of the normal wing weight whereas, if the guns had been installed within the wing, not only would this weight penalty have been paid but additional weight cost would have been incurred due to the complication of the structure.

Fuselage Design

The structural diagram of the L-200-1 fuselage is shown on pp. 42-43. Here again a major simplification was achieved because of the nature of the airplane and because of the incorporation of a gear box with a large offset. The major structure consisted of four main longerons which were located well to the top and well to the bottom of the fuselage, thereby achieving maximum structural strength and stiffness for a minimum weight. Attached to these longerons were several main bulkhead rings, all of which had a circular cross section. This further simplified the production of the airplane. Provisions for the power plant were made by providing engine mounts on two of the major bulkheads. Since the power plant was located outside of the major fuselage structure, the provision for tail pipe heat and for the installation of the tail pipes themselves were extremely simple and consisted only of external shrouds attached to the normal fuselage structure. By separating the oil coolers, the ducts for the oil coolers were quite small and did not cause any serious structural difficulty. For the main engine inlet, the duct was directly forward of the engine and occurred between the two lower longerons where it could be easily re-

38) **The L-200 wing had a much simpler structure than was typical on high performance airplanes. Only one control surface was used—the aileron-flap combination—and it was hinged at five points.**

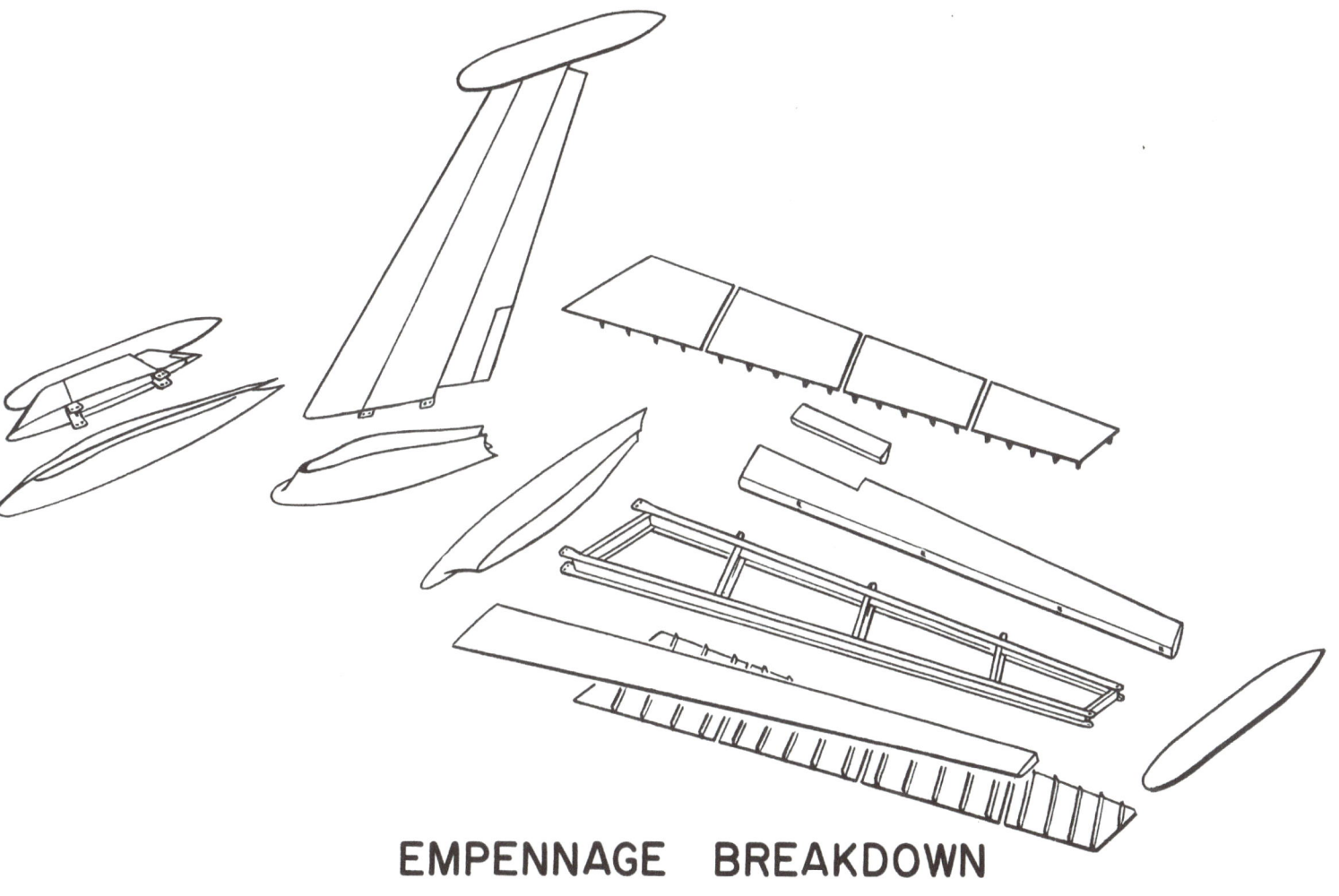

EMPENNAGE BREAKDOWN

▲ 39

moved and where no additional penalty was paid since a cutout was already made for the engine installation itself.

Titanium was used in the fuselage structure which was also used for fire protection. This included a floor over the engine and splitters between engines.

Empennage Design

The drawing on pp. 44-45 illustrates the one major structural difficulty which was incurred in an airplane of this type since in the empennage not only did the air loads have to be withstood but also the landing loads. This required a concentrated effort to reduce weight and Lockheed proposed that this be done by the utilization of titanium forged empennage spars and one of the aft fuselage attachment bulkheads. These parts appeared to be reasonable to build in production and promised a major saving in structural weight. For prototype work it was proposed that an effort be made to obtain these titanium forgings if the program accepted by the Navy included the full-scale prototype airplanes. If the small-scale prototype was utilized, it was very probable that machined dural spars or built-up steel spars could be incorporated with a weight penalty.

The use of extruded integrally stiffened skin panels was proposed for the fixed surface structure. Experiments and production use of this technique had shown that a considerable weight saving could be made along with a reduction in cost. A further structural innovation was shown in the elevator construction where a simple shell was filled with Lockfoam plastic. Production parts installed on F-94 all-weather fighters had demonstrated the great value of this type of structure in strengthening and stiffening control surfaces.

39) The L-200 empennage consisted of a fin and two stabilizers. Each section was interchangeable and the identical stabilizer assembly was used on each side.

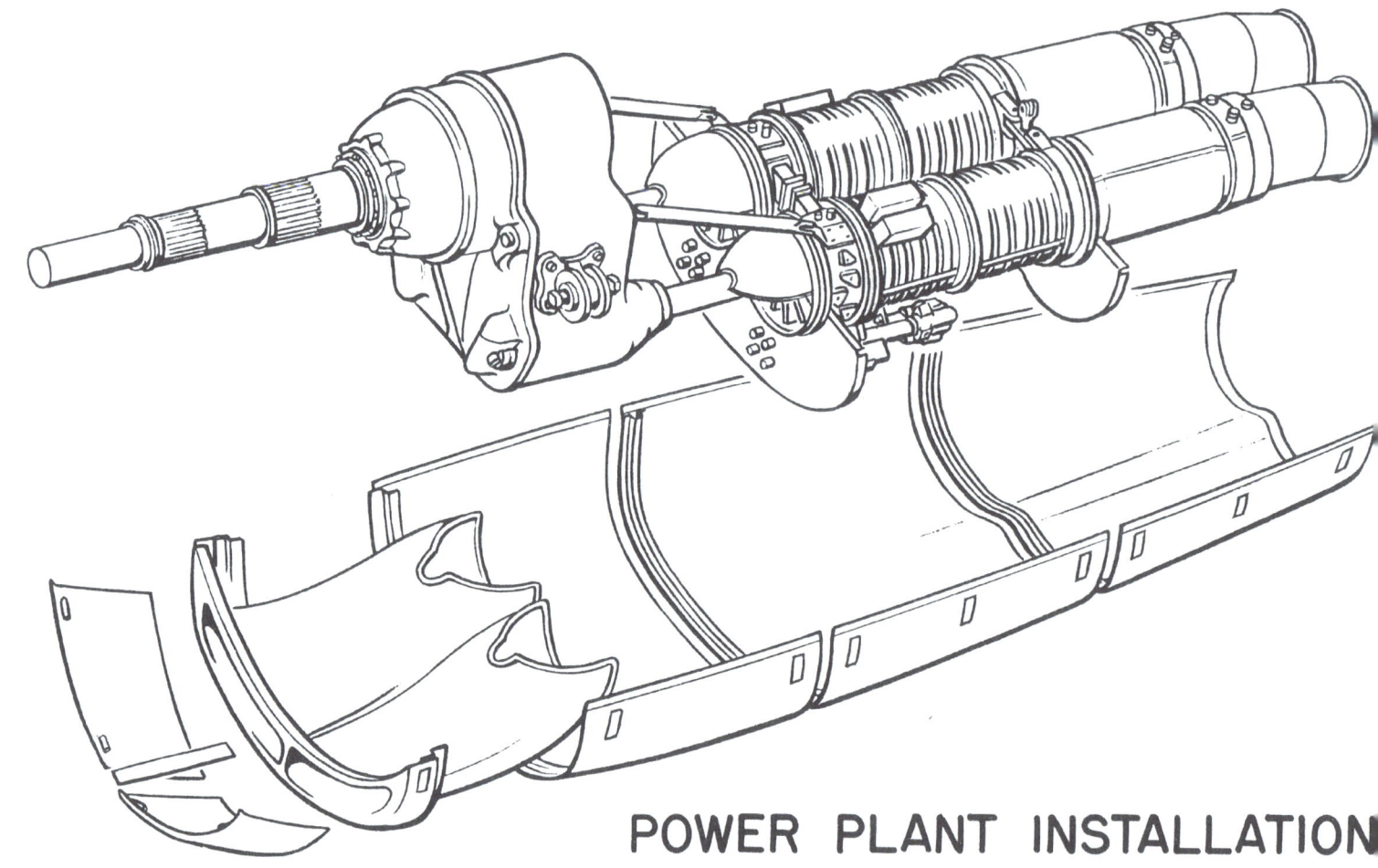

POWER PLANT INSTALLATION

▲ 40
L-200 Production Features

While the L-200 was presented as a prototype airplane, considerable effort was expended to assure that it could be converted into a production model in a minimum amount of time. The same assembly tooling used for the full-scale prototype could in some cases be used for the production version. And, as shown in the production breakdown drawing on pp. 46-47, the airplane was well adapted to a very high production rate. Nearly all the prototype layouts would be used unchanged on the production airplane; the only additions, other than changes resulting from flight test, would be the procurement of forgings and extrusions that would be uneconomical to use in small quantities.

Lockheed pointed out that its plan for a lightweight full-sized prototype would further reduce the span between the time when the prototype order was placed and the first production airplane was delivered. This was due not only to savings that could be made in design and flight test time, but to the fact that manufacturing pre-planning and tool design could be started early in the prototype design stage. This procedure, which was first started at Lockheed with the R6O *Constitution*, had been made standard to reduce the gap between the prototype and production airplanes.

Production Breakdown

The production breakdown was identical for both the full-sized prototype and production airplanes, and was based on service, spare part requirements and producibility. For the first consideration, the major airplane components that were subject to damage or wear were provided with simple interchangeable joints requiring a minimum number of attachments. These components were:
1. Complete Wing
2. Complete Empennage

40) The engines, engine accessories, and gear box of the L-200 were accessible for repair or removal through the bottom of the fuselage. The cowling was removed in three sections, and the intake ducts were removed as a unit.

3. Fin
4. Stabilizer
5. Rudder
6. Elevator
7. Landing Pods
8. Armament Pods
9. Ailerons
10. Engine Fairing
11. Canopy
12. Power Plant
13. Electronic Equipment Package
14. Radome

Airplane components that were readily replaceable by simple drilling or trimming operations were:

1. Wing Leading Edge
2. Wing Trailing Edge

Shipping

As shipping was considered an important factor, the airplane was designed so any component would fit in the 30' x 9' x 7' crate noted in the crating specification AN-C-118a. In order to accomplish this, the wing was separated from the fuselage by removing the attaching bolts and the wing leading and trailing edge assemblies which were attached with a series of screws. The fin, stabilizer, propellers, and radome were removed and the fuselage was then ready for crating. This crating arrangement allowed transport by either rail or ship, and no special routing or handling equipment was necessary.

Producibility

The consideration of producibility which further influenced the design was made up of three main factors:

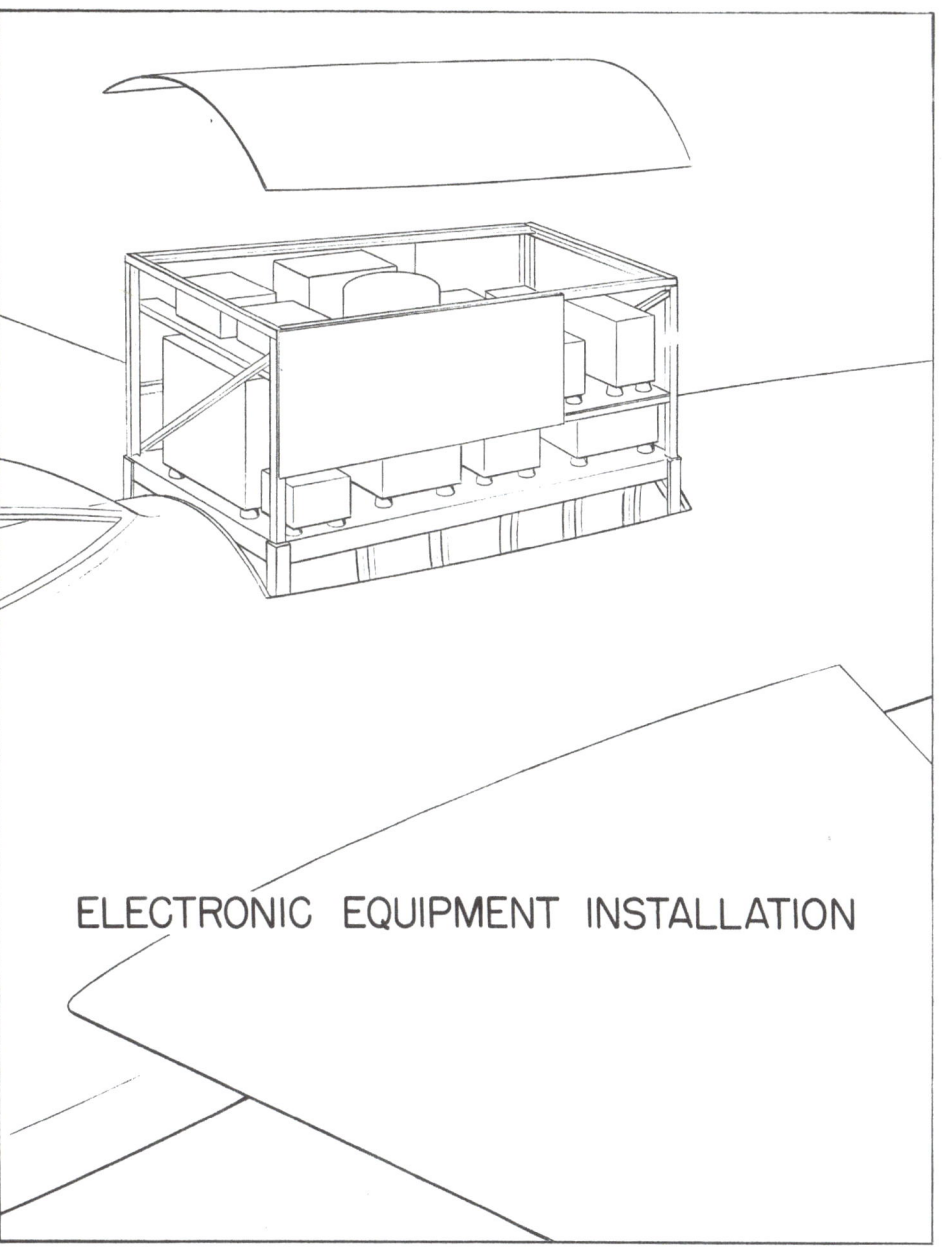

▲ 41

ELECTRONIC EQUIPMENT INSTALLATION

41) To reduce the maintenance-to-flight time ratio, most of the electronic equipment was installed in a compact rack in the aft section of the fuselage. The rack could be raised for checkout and alignment of the individual units, or could be removed entirely.

Minimum Man-Hours

In order to accomplish this, the number of parts in the airplane was kept to a minimum. An example was the use of integrally stiffened skin for the wing surface structure. This type of structure not only reduced the number of parts (by eliminating separate stiffening members), but also eliminated a large number of attachments; it was proven in actual practice at Lockheed to be a means of saving a large number of man-hours. Man-hours were also saved by designing for assembly access. The fuselage, for example, was made in half-shells which not only allowed unrestricted access for workmen during fabrication, but also permitted installation of part of the equipment before the shells were joined.

Economy in Tooling and Floor Space

The use of simple joints, such as that of the empennage forged bulkhead to the fuselage, reduced the complexity of tooling and, in more complicated joints such as the four-bolt fin attachment, adjustment was designed into the forward beam fitting so it could be readily aligned with the rear beam fitting. The fairing between the fuselage and fin attached to the fuselage, which further simplified the mating problem.

Tooling and floor space were both saved by multiple use of assemblies. This was done by designing the part so it could be used on both right and left-hand sides of the airplane. Major examples of this were:

1. Wing Leading Edge Assemblies
2. Wing Trailing Edge Assemblies
3. Ailerons
4. Horizontal Stabilizers
5. Elevators
6. Landing Pods

Subcontracting and Expandability

The airplane breakdown was well adapted to subcontracting and expanding production at a mobilization rate. The main assemblies were small and, therefore, easily handled and transported. They were complete units with simple joints. This arrangement allowed short final assembly lines which could be established at several different airports with very low tooling duplication.

The manufacturing techniques used in the production of the L-200 were standard methods of manufacture and were all used on contemporary Lockheed production airplanes.

Critical Materials

The use of the more critical materials was avoided as far as possible in the L-200 design. Titanium was used instead of stainless steel as it was expected that this material would be in better supply than stainless when the airplane went into production. In the event that titanium was not available, aluminum-coated low carbon steel would be used in place of titanium sheet and 8630 steel forgings used for the empennage bulkheads and beams.

Production Research

Various research projects were underway at Lockheed to develop new types of structure which would result in either weight or cost savings. Among those that were possibly applicable to this airplane were: large thin-walled castings that could be used for leading edges and Lockfoam sandwich panels that could be used for wing surface structure. Lockfoam was a Lockheed developed, low density, poured-in-place structural foam that was used as a filler to eliminate ribs and stringers in structures such as control surfaces. In this application it had been thoroughly tested and was used in contemporary production airplanes. In addition, experimental designs utilizing titanium were in the process of construction. Large thin web forgings for wing surface structure were under development.

Fuselage

The fuselage, as shown on p. 48, broke down into five major assemblies: left and right side panels, floor and engine mount sections, lower aft section, and empennage attaching section.

This type of breakdown was advantageous as it produced sections that were easily handled, had good access for assembly, broke along natural lines (requiring no extra joints), and allowed installation of equipment in the subassemblies. The mating operation consisted of joining the first four assemblies and drilling the fittings for attachment of the empennage section. One of the primary aims in determining the breakdown was to keep as much work as possible in the subassemblies, thus relieving the congestion around the complete fuselage.

The fuel tank and cockpit floor assembly contained the engine mounts which were completely installed in this assembly. The lower aft section was titanium and was a spotwelded assembly.

The empennage attaching section connected to the forward section of the fuselage with six tension bolts. A joint of this type could be made interchangeable with very simple tooling; this was a desirable feature since, in the event of a subcontracting program, the empennage attaching section could be built by the same manufacturer that constructed the fin and stabilizer. The more difficult problem of coordinating the fin and stabilizer joints with the empennage attaching section was thus confined to one manufacturer.

Wing

The basic design features of the L-200 wing produced a much simpler structure than was usual on high performance airplanes. Only one control surface was used—the aileron-flap combination—and it was hinged at five points. As all the fuel was carried in the fuselage, the wing was relieved of all complexities resulting from its use as a fuel carrier. The ammunition was carried on either side of the main beam web and was accessible from the wing tips or by removing the fore and aft surface structure.

The wing as a complete unit was interchangeable and was attached to the fuselage by five bolts on each side. The nose section and the aft section were attached to the main beam with screws and plate nuts and were removable or replaceable for quick repair of battle damage. The wing was assembled to the fuselage by removing the nose and aft sections and slid in the main beam through the slot in the fuselage. The fore and aft assemblies were then joined to the beam with screws, the aileron was installed, equipment connections made, and the wing was ready for service.

The main beam consisted of a web and two extrusions that were continuous from tip to tip. The extrusions were well within the limits of standard extrusion equipment. The fore and aft sections were beam and rib structures with extruded integrally stiffened skin which not only reduced the number of parts and attachments in the wing, but also allowed access in subassembly.

The aileron was a Lockfoam-filled type which was used on contemporary Lockheed production airplanes.

The armament pod was joined to the wing with three bolts accessible through the door in the pod. The joint was interchangeable to allow rapid replacement of either pod.

Empennage

The empennage consisted of a fin and two stabilizers. Each section was interchangeable and the identical stabilizer assembly was used on each side. The empennage sections attached to the fuselage with four bolts each, and the fairing, which covered the joints, attached to the fuselage only.

The structure of the stabilizer, which was typical for the fin, consisted of two forged titanium or steel beams separated by five aluminum alloy ribs. The skin was extruded integrally stiffened aluminum alloy which substantially reduced the number of ribs and attachments. The nose section was a separate assembly built of ribs and skin. It was attached to the front beam with blind fastenings or screws.

The elevators and rudder were Lockfoam-filled which eliminated a large number of parts. The hinges had adjustable brackets to ensure alignment, and the inner bracket took all the end load, allowing clearance between the remaining fittings.

The landing pods attached to the ends of the beams with two bolts; the identical pod assembly was used in all three positions.

Power Plant Installation

The engines, engine accessories, and gear box were accessible for repair or removal through the bottom of the fuselage. The cowling was removed in three sections, and the intake ducts were removed as a unit. The duct closure doors could be either opened on hinges or removed for better access.

In production, it was expected that the engine and gear box would be installed as a unit. The cowling and ducts were aluminum alloy and were attached with quick-acting cowl fasteners.

The gear box and power units were isolated from the upper fuselage by means of titanium fire barriers. The engine compressor section between the compressor inlet and burners was also compartmentalized in this manner. All potential fire compartments were adequately vented.

Electronic Equipment

Experience gained on the F-94, the only contemporary automatic tracking radar-equipped interceptor that was delivered in production quantities, had demonstrated the importance of design in reducing the maintenance-to-flight time ratio.

To accomplish this, as much of the electronic equipment as possible was installed in a compact rack in the aft section of the fuselage. The rack could be raised for checkout and alignment of the individual units, or could be removed entirely. In production this entire unit would be a bench assembly and, to a large extent, could be checked out before installation in the airplane.

The radome was attached by two quick-acting fasteners that were accessible from the outside.

Weight and Balance Analysis

Since the success of the vertical rising airplane depended not only on the aerodynamic problems of flight but also on the engineering ability to build a satisfactorily strong airplane for minimum weight, it was decided that drastic procedures to hold weight down were justified. Accordingly, Lockheed suggested the following procedure to ensure that minimum structural weight was achieved:

1. Design and construct two T-40 powered full-scale prototypes in place of the small Mamba powered airplanes. These prototypes were to be designed for a low weight corresponding to their use as prototypes (11,578 lbs).
2. Construct one static test article identical to prototypes, including low design weight.
3. By means of flight test load determination on prototypes and application of these loads to the static test article, a structural beef-up program was instigated to increase the design gross weight to the tactical requirements.

Following this procedure, the first tactical airplane prototype (a converted and beefed-up original prototype) would represent the lightest achievable vehicle, an essentially "stretched" airplane which would compete in weight efficiency with airplanes which had years of operational stretch to achieve their actual operational gross weights.

It was the aim of the above program outline to have the prototype airplane be the zero point for the L-200 airplane and have the weight of the tactical airplane determined by the weight trend of previous Lockheed aircraft. This, of course, applied only to the structural components of the airplane, the added equipment, changed power plant, armament, etc. having been accounted for by direct addition. Thus, the weight empty of the tactical airplane could be determined as follows:

9,435	Weight empty of prototype with T-40A-6 engine.
757	Weight of added equipment: radar, extra fuel capacity, etc.
1,022	Weight of changing T-40A-6 to T-40A-8 engine, including propeller change.
101	Increase in structural weight from weight history curve.
11,315	Total weight empty of tactical airplane (lbs)

To illustrate the possible weight saved by this program, the weight empty of the airplane was estimated by normal procedures similar to those used in estimating the original prototype weights. These calculations showed:

12,416	Weight based on conventional estimating procedures.
11,315	Weight empty developed through growth program.
1,101	Savings through growth program (lbs)

The table at the top of p. 49 is a breakdown of all the required weights for performance calculations based on the recommendations and definitions of 0S-122. Below this is a complete summary of the weight empty breakdown estimate based on the suggested prototype program.

Of major interest to those who expected a swept wing on airplanes with performance approaching the transonic, Lockheed emphasized that an extremely thin wing was determined to be lighter than a swept wing for a given drag and drag rise Mach number. A direct comparison between the straight wing and the swept wing showed the straight wing to save 210 pounds or over 21% of the total wing weight.

Conclusions

It was the conclusion of the Lockheed Corporation from these preliminary design studies that a feasible and useful tactical airplane could be developed to fulfill Navy Specification 0S-122. Lockheed believed their design to have the best chance of fulfilling these requirements satisfactorily. However, the company emphasized that problems of this nature required a major research effort. Fulfillment of all of the requirements which could be uncovered in attempting vertical flight, and the transition between vertical and horizontal flight, could result in major design changes.

It was concluded from this study that the best type of research program to obtain the most desirable end result would include a prototype airplane that was full-scale rather than small-scale as proposed in the original Navy request for proposal. Reasons were given for this suggested change in the program, but it was concluded that the airplane could be developed either with a small-scale prototype or with a large-scale prototype, depending upon the desires of the Navy after studying the problem as stated in the proposal.

With respect to the performance requirements of the airplane, it appeared reasonable to achieve the combat altitude and speeds required; it appeared entirely feasible to make deck facilities which would ensure reasonably normal operations for takeoff and landing on a commercial vessel; and it appeared entirely possible to develop an automatic control from an existing autopilot which would relieve the pilot of many responsibilities that were of an unconventional nature. It was the major conclusion of this study that the unconventional performance requirements could be achieved with a conventional airplane, thereby relieving the research worker and the eventual pilot of the airplane of many incidental problems which were always a major stumbling block to progress.

Lockheed believed that any research programs which would lead to the development of an airplane of this nature were extremely valuable to the progress of Naval aviation since a successful vertical rising airplane would open up many fields in which the airplane would become even more useful from a tactical standpoint than it had been in the past. These fields included ASW work, mining operations, reconnaissance, attack, and even advanced utility transports.

The dimensions of the L-200 were limited in such a manner that the airplane on its service cart would fit on any of the carrier elevators previously listed by the Navy for the carrier-based ASW airplane. The included the CVB-41, CV-9, CV-34, CVL-48, CVL-22 and CVE-105 carriers. Further efforts would be made in the first phases of the L-200 study to reduce the size of the airplane further by concentrated wind tunnel research programs to determine minimum possible tail lengths. It was hoped that this would make possible a total airplane length below 41 ft so that small carriers (then inactivated) could also be used to transport these airplanes or serve as a service ship.

Cockpit Mock-up

In addition to the L-200 design summary covered in the previous section, Lockheed also submitted a report to BuAer concerning its study of a cockpit mock-up for the aircraft. According to the document, although considerable research had been made on the problem of pilot position in vertical flight at NADC in Johnsville, it was considered desirable to construct a simple cockpit mock-up for the Lockheed design proposal. In order to gain a better understanding of the movable seat problem with relation to visibility requirements, the Mark VI gun sight installation, ejectable seat clearances, and other equipment items, a full-scale wooden mock-up of the L-200-1 cockpit was constructed. It was designed to pivot in a supporting frame from a normal to a vertical position, power actuated to simulate transition in flight, and equipped with a tiltable, rotatable seat.

The tilting seat was decided upon for use in the airplane after mock-up inspection verified the advantages of this method over other proposed methods of positioning the pilot for vertical flight.

The mock-up permitted three-dimensional study of the various cockpit arrangement problems presented by the limited space envelope and the necessity of maintaining satisfactory control motions and visibility in all seat positions. The resultant arrangement appeared to be satisfactory in all respects, although compromises were necessary in resolving the prob-

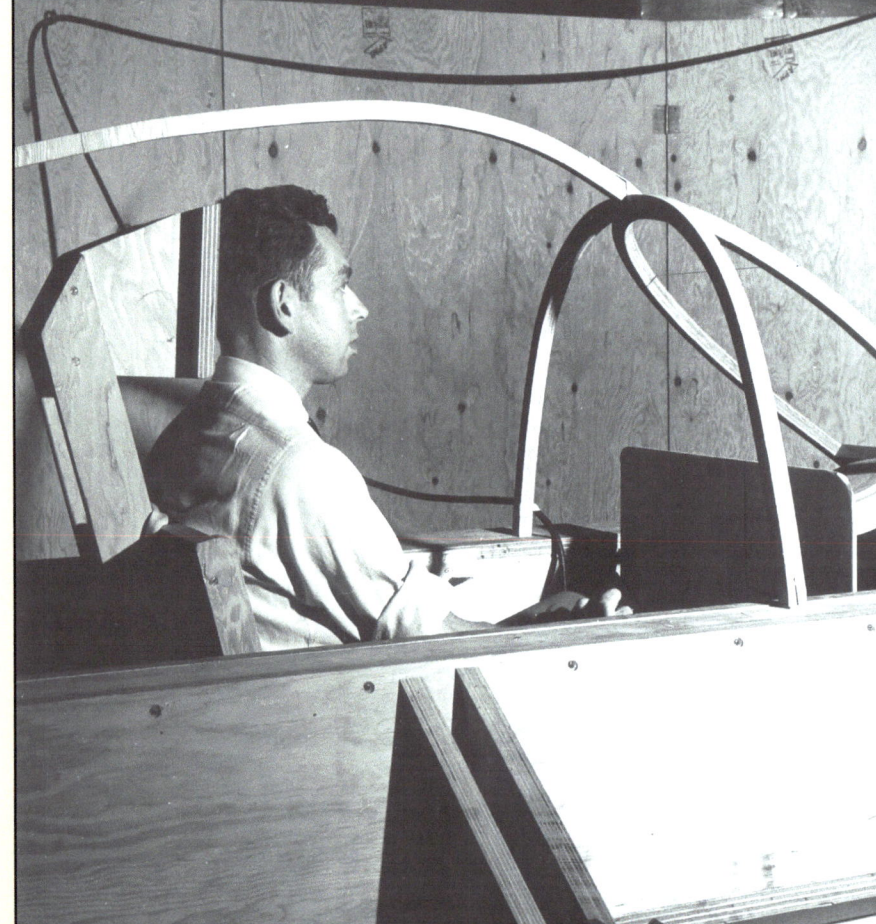

42) Photo of the plywood L-200 cockpit mock-up, constructed on Lockheed's initiative to investigate the ideal position for the pilot.

43) The mock-up in horizontal flight attitude with normal seat position.

▲ 44

lems imposed by the unconventional features with the requirements of applicable specifications.

44) The mock-up could rotate from the normal position to a vertical attitude by means of an electric winch, simulating to some extent the sensation of flight transition.

Introduction

The construction of a full-scale wooden mock-up of the L-200-1 cockpit was prompted by the necessity of evaluating the relative merits of the various methods of moving the pilot from his normal position to a position suitable for the

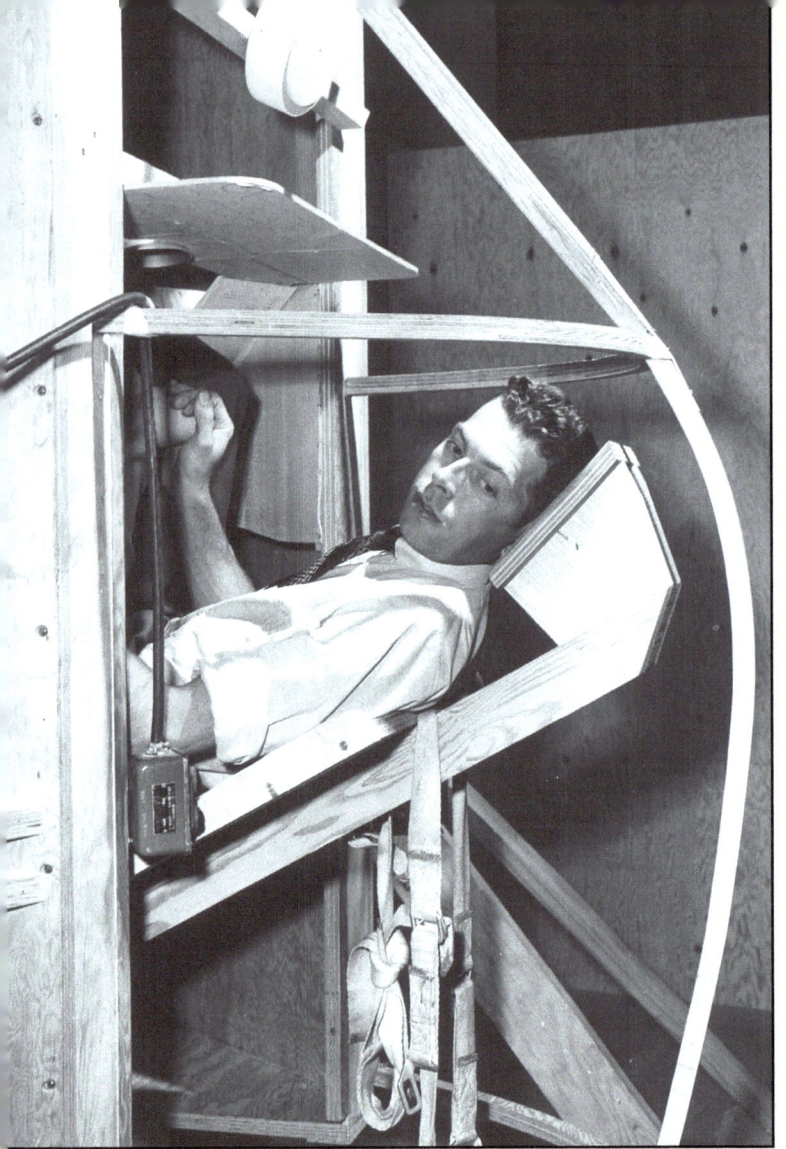
▲ 45

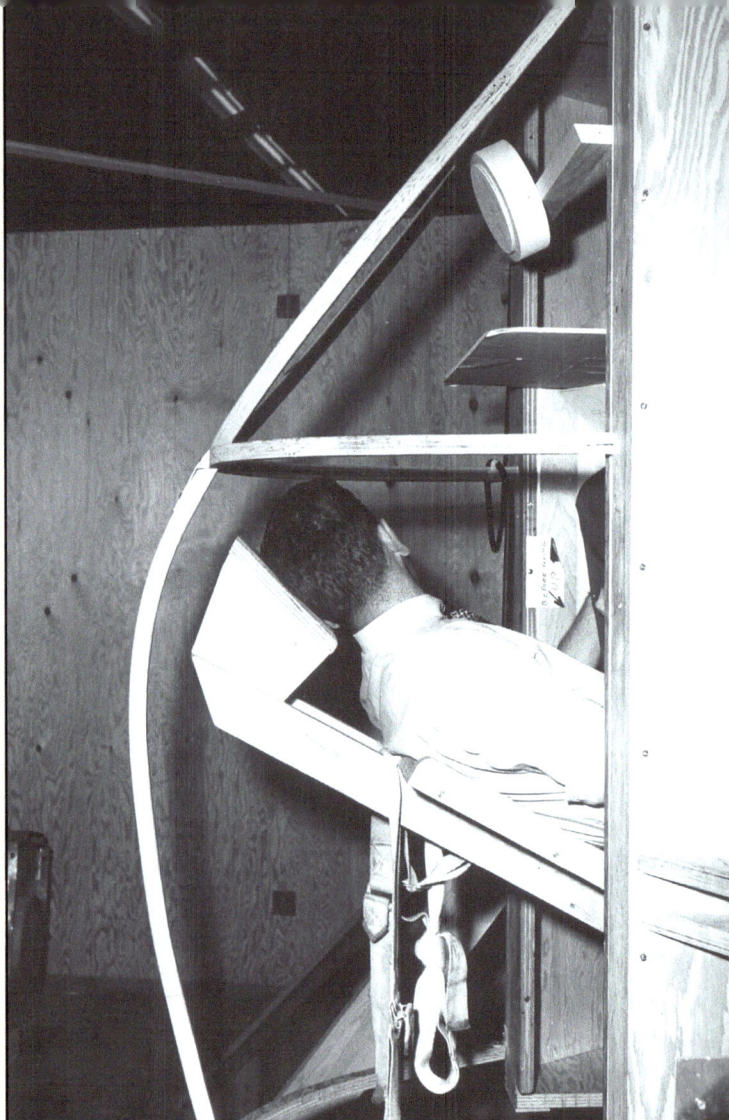
▲ 46

vertical flight attitude. Details of the mock-up design are described in Section I.

As a result of the mock-up program, it was decided that the tilting of the pilot's seat forward about a horizontal axis was the most practical method. Section II on p. 62 is a discussion of the factors which led to that decision.

A cockpit arrangement based on the tilting seat design was studied in detail, and the mock-up proved to be very helpful in arriving at practical solutions to the problems presented by the unconventional features of the L-200. A description of the cockpit arrangement and a discussion of some of the major problems appear in Section III on p. 64.

The mock-up was also used to demon-strate the practicability of entrance to and egress from the cockpit in the vertical attitude with the seat tilted back to its normal position.

Section I— Mock-up Design

The L-200 cockpit mock-up, constructed mainly of 1" plywood, was designed to approximate the airplane cross section above the floor line, using straight sloping sides and a constant width to facilitate construction. The dimensions of the box were 76" long by 66" wide at the floor line. A 1 x 1 canopy and windshield frame was included, and the complete cockpit was supported in a heavy wooden A-frame on a horizontal steel-tube pivot axis located approximately at the center of gravity of the occupied cockpit. Stops were provided at both horizontal and vertical positions.

The pilot's seat was designed to rotate on a turntable about a vertical axis as well as to

45-46) These photos show the mock-up in hovering or takeoff attitude with the seat tilted 45°; the pilot stand-in is demonstrating the proposed method of over-shoulder sighting during landing operations.

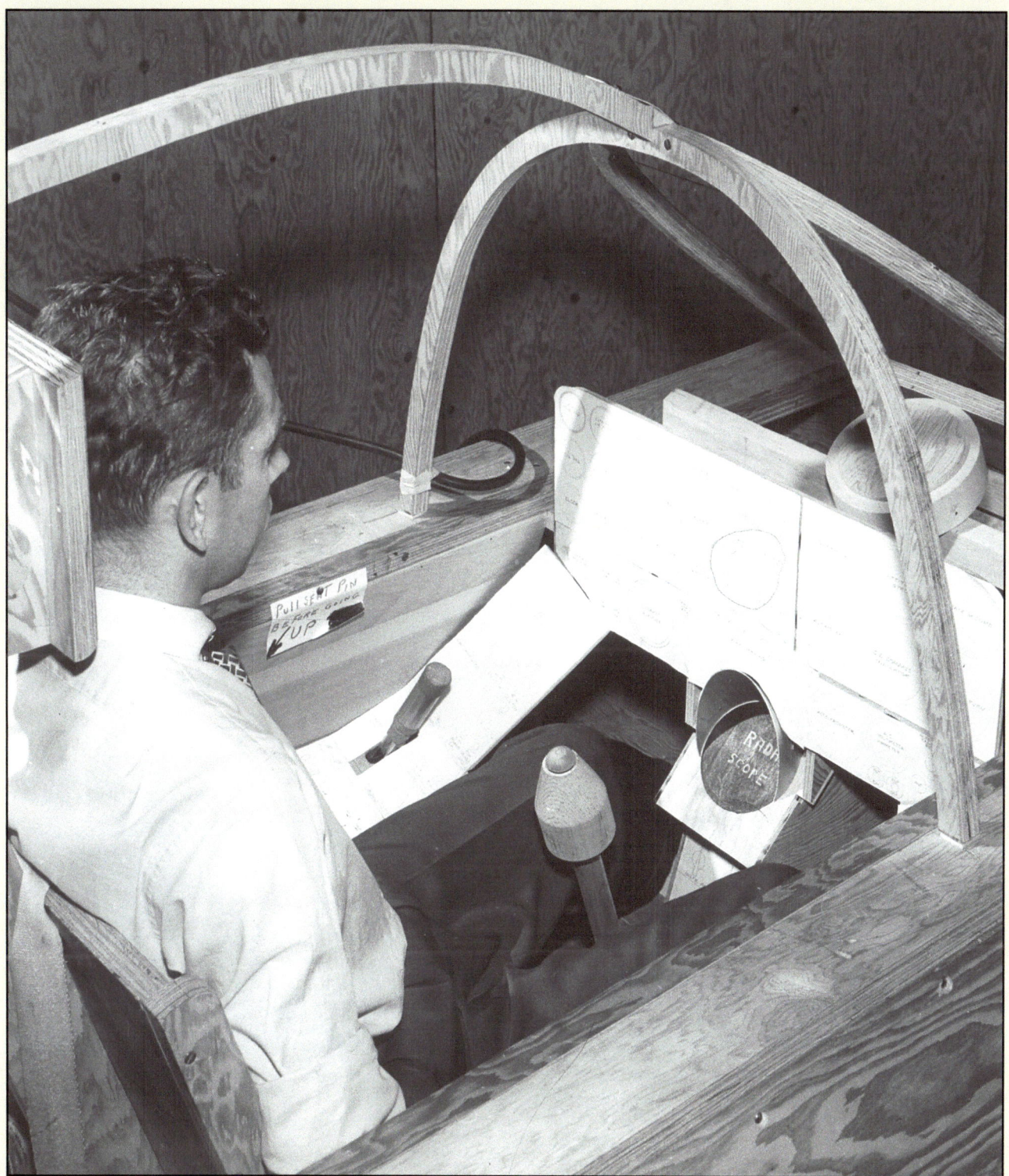

▲ 47

tilt about a horizontal axis. The turntable was pinned in a fixed position when it was decided to abandon the rotational ideas in favor of the tilting seat.

The seat, canopy, floor line, cowl line, instrument panel, control stick, rudder pedals, and other items were corrected frequently as the airplane design progressed, and the final mock-up configuration as pictured in Photos 42-51 differed from the proposal drawings only in minor details.

In order to simulate to some extent the sensation of transition, an electric winch was rigged up to hoist the cockpit from its normal position to a vertical attitude (Photo 44). The winch was controlled by a switch box attached to the cockpit, permitting a person sitting in the pi-

47) The hooded radar scope was located below the instrument panel with the sight head face forward of the panel.

lot's seat to fully control the transition. The tilting of the seat was automatically synchronized with the cockpit motion by means of a cable attached to the seat at one end and to the frame at the other, passing over pulleys mounted on the cockpit box.

Section II— Pilot's Tilting Seat

Prior to construction of the mock-up, many different methods of changing the pilot's position were considered. The use of a prone position in normal flight was discarded because it was felt that the unconventional features of this airplane should not be augmented by the addition of an unconventional normal flight position.

After preliminary evaluation of remaining methods with regard to comfort, visibility, safety, and space requirements, all others were discarded in favor of a choice between rotation about

▲ 48 ▼ 49

48) The pilot had to lean forward in order to cover the full range of gunsight lead angles without head movement.

49) The left-hand console, included the APQ-42 control panel with its tracking lever, the trim tab control switch, the power control lever and switches for various engine control functions.

▲ 50 ▼ 51

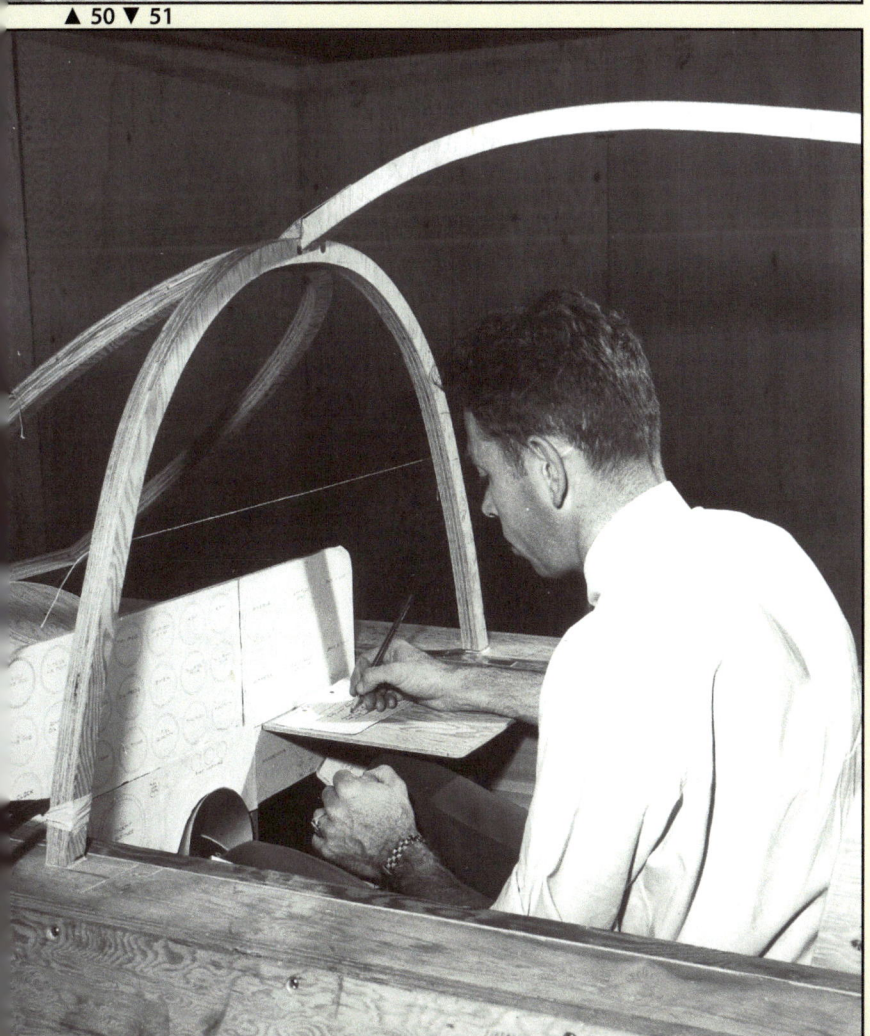

a vertical or inclined axis and tilting about a horizontal axis, as tested in the Johnsville mock-up. Inasmuch as the lateral space requirements of an inclined axis of rotation were incompatible with an aerodynamically efficient cross section, the mock-up was designed to include only provisions for seat movement about horizontal and vertical axes.

From inspection of the completed mock-up, it was readily apparent that rotation about a vertical axis, while possible within the chosen cross section, eliminated the use of considerable valuable space surrounding the seat. Furthermore, it would still be necessary to tilt the seat after rotation in order to achieve the desired position. Since the desired position was approximately one in which the seat and back locations were interchanged, consideration was given to the possibility of having the pilot change position without moving the seat. This idea was also abandoned as impractical and unsafe in such cramped quarters.

The tilting sent idea was adopted as the most practical because of the following advantages:
1. Hands and feet could remain on controls during entire transition.
2. Duplication of controls and instruments was unnecessary.
3. Minimum space requirements.
4. Satisfactory visibility and comfort in hovering attitude.

The location of the tilt axis was dictated by the necessity of retain-

50) Photo of the right-hand console with the W-2 autopilot controller located aft of the console control panels.

51) The chart board/writing pad holder in the extended position.

ing satisfactory control stick and rudder pedal movement and instrument visibility, plus the clearance requirements of the canopy, which was already limited by the gunsight requirements on one hand and aerodynamic efficiency on the other. The proposed location of the tilt axis was 6.5" above and 6.5" forward of the seat reference point in mean position. A more forward location of the seat pivot, as used in the Johnsville mock-up, necessitated either a greater distance to the 35° windshield or a lower normal position in order to provide head clearance in the tilted position, resulting in sacrifice of specified visibility.

A tilt angle of 45° from the normal position was adopted as a reasonable compromise between comfort and visibility. In order to minimize the tilted position clearance problem, it was decided to provide for tilting only from the low position of the seat. This required the pilot to lower the seat prior to tilting during transition, a motion which could be utilized to arm the electrical circuit controlling the proposed automatic tilting procedure during transition. Since the low position was also a prerequisite for seat ejection, the seat could be readily positioned for ejection from either attitude.

Because the cockpit was sandwiched between a floor level fixed by the entire gearbox and a canopy whose height was limited by aerodynamic requirements, it was decided to limit the vertical adjustment of the seat to 2" up or down from the mean position, along a line sloping 23½° forward. This adjustment moved the seat within a frame supported at the tilt axis by a carriage which in turn rode on rollers in the ejection rails. The heel rests for ejection were attached to the carriage instead of to the tiltable portion of the seat in order to minimize floor clearance problems.

Photo 43 shows the mock-up in horizontal flight attitude with normal seat position. The hovering or takeoff attitude, with seat tilted 45°, is pictured in Photos 45-46. These photographs indicate the proposed method of over-shoulder sighting during landing operations. A mirror attached to the canopy frame was also considered and could be used to advantage to relieve fatigue induced by prolonged hovering.

Section III—Cockpit Arrangement

The mock-up was used to good advantage in establishing the proposed cockpit arrangement. The limitations imposed by the visibility requirements over the large diameter spinner, the presence of the engine gearbox below the pressurized floor, the windshield slope as dictated by the optical sight head, and the ejection seat clearance requirements all had to be correlated with the tilting seat and the attendant control motion problems.

Because of the gearbox below the floor, the control stick was designed to pivot about a horizontal lateral torque tube just above the floor, with a pivot for lateral stick movement located 7" above the torque tube. By limiting the stick height to 20" above the torque tube, it was possible to use 14" fore-and-aft stick travel and 12" total lateral travel in either the normal or tilted position of the seat.

After consideration of normally hung rudder pedals and a mock-up tryout of a rudder bar, the use of sliding pedals was determined to be the most adaptable to the change of seat position. Pedals were inclined at a 45° angle and provided with toe stirrups to help retain the feet in position during hovering.

One of the necessary compromises worked out on the mock-up was the centerline location of both the Mark 6 Mod 1 sight head and the APQ-42 radar indicator, necessitating the offset of the primary flight instrument group to the right because of insufficient space between the sight head and the instrument panel, the location of which was limited by ejection clearance requirements and visibility specifications. Photo 47 shows the location of the hooded radar scope below the instrument panel and the sight head face forward of the panel.

In order to cover the full range of gunsight lead angles without head movement, it was necessary for the pilot to lean forward to the position shown in Photo 48.

Maximum utilization of available space was obtained by sloping the side consoles for optimum visibility and continuing the consoles forward on an upward slope to the corners of the instrument panel. The left-hand console, pictured in Photo 49, included the APQ-42 control

▲ 52

panel with its tracking lever, the trim tab control switch, the power control lever and switches for various engine control functions. The landing gear switch was located at the top of the sloping panel, flanked by the APN-22 radar altimeter indicator located for ready reference in the hovering attitude. The engine instruments were placed in a compact group to the left of the primary flight instruments. Also visible was the central switch panel for interior and exterior lights and the fuel system diagram, which included switches for the four fuel shut-off valves.

The oxygen regulator was located on the vertical panel below the sloping side panel. The right-hand console is shown in Photo 50, with the W-2 autopilot controller located aft of the console control panels, which included the ARC-27, ARR-2A and APX-6 controls. The upward sloping panel included seat and canopy switches, cabin temperature controls, oil cooler switches, starter switch, DC power controls and "essential bus" switches. AC power controls were on the lower portion of the vertical panel below the flight instrument group.

The autopilot controller mentioned above was intended to be used only during normal flight operations; during hovering, the psychological advantage of manual stick operation could be retained by the use of a four-way momentary contact switch on the top of the control stick to effect required changes in the hovering attitude while stabilized by the autopilot. This allowed the pilot to change from automatic flight to manual flight without removing his hands from the stick. The switch button and the switch housing are shown on the stick in Photos 49-50.

A cylindrical knob for simultaneous adjustment of both rudder pedals was located on the left-hand side of the control console, and is visible in Photo 50.

The slot below the flight instrument group is the stowed position of a 9" x 12" chart board or writing pad holder, which is shown in the extended position in Photo 51.

52) **Original display model of the Lockheed L-200 Convoy Fighter, posed here as if in flight. Note the representation of the retractable "fingers" on the tail pods, which were to be deployed when the aircraft was recovered on the shipboard landing net.** *(John Aldaz Collection)*

Free Flight Model Tests

Tests on a free flight model (unrestrained in pitch) were made as a part of the Lockheed design proposal for the vertical-rising Convoy Fighter. The purpose was to study model and airplane behavior in hovering flight, ground or deck effects on control, airplane attitude with translational or side-wind velocity, and transition flight paths. Also included were in-place stability on the landing-takeoff platform and actual takeoffs and landings from the floor and the platform.

Two models were used in this program; the first a conventional commercial kit and the second was a 1/16 scale model of the actual configuration proposed for the design competition. Both were powered with internal combustion model engines. The sixteenth scale factor was determined primarily from scale propeller diameter and model engine availability.

It was determined as a result of the tests that hovering control of the model was very good and that the full scale airplane would be even better. Manual flight of the airplane in the event of auto pilot failure would be readily achieved. Good control of the airplane and model was possible with ground or deck close to the tail. Transitions from vertical to horizontal were made with minimum attainment of altitude.

Results of tests to determine pitch attitude versus translational speed of the model were checked very closely by wind tunnel measurements.

Introduction

At the inception of the design of the L-200 vertical-rising Convoy Fighter, the stability and control of the aircraft in hovering flight was a major problem. Free flight tests conducted by NACA with radio-controlled tail surfaces indicated instability or divergence with long periods which appeared to be easily controlled by the operator. Moving pictures of these tests were encouraging with respect to pilot control when hovering in what appeared to be calm air and at an appreciable distance from the ground. However, ground plane effects and various types of disturbances from retrieving and launching devices were still an unknown factor.

The construction and testing of a free flight model including the evaluation of side wind velocities, retrieving and launching gear effects, in addition to hovering, and transitional flight appeared to be very desirable. Dual rotation propellers and radio-operated control surfaces were considered for a scale model but were abandoned because the cost and time required for development were prohibitive for a design proposal.

Models powered with small internal combustion engines and controlled through fine diameter flying wires offered a solution to the time and expense problem.

In addition, it was known that the low disc loading of the propellers of model airplanes were favorable for hovering and vertical climbing flight. Satisfactory thrust-weight ratios could be achieved to study some of the problems of the full scale airplane. Further, this type of model flown by an operator with a "U-control" handle which operated the elevators, was completely unrestrained about the pitch axis and offered opportunities for studying many of the problems of the full scale airplane. With a slight modification, data could have been obtained in the yaw direction although this was not done in the tests described in the report due to lack of time. The model was necessarily restrained about the rolling axis because of the single rotation propeller and its resultant torque.

Having decided on a model incorporating an internal combustion engine, construction was started on a standard kit model which was available commercially. The problems of fuel feed in vertical flight and thrust control by the operator were solved with commercially available items. Meanwhile, the design of the full scale airplane in the Preliminary Design group progressed to a point where the general external configuration was firm. A 1/16 scale model of this configuration was constructed using the same type of fuel feed and power control. Preliminary tests were conducted on the conventional model, followed

---

53-56) Multiple views of the 1/16 scale free flight model of the L-200 Convoy Fighter. This represents an earlier configuration of the design, with the empennage rotated 180° and no wing dihedral.

▲ 53 ▼ 54   ▲ 55 ▼ 56

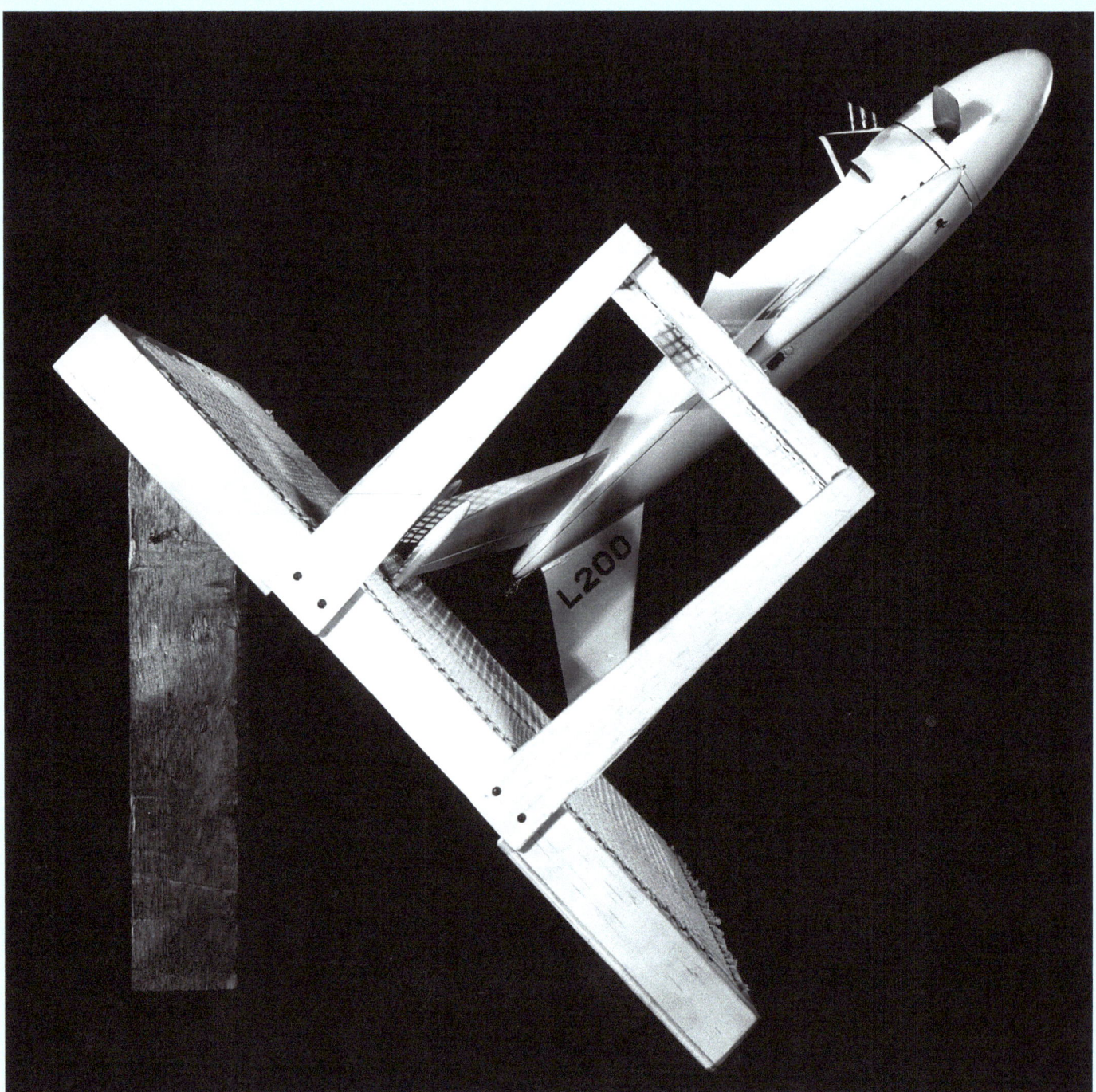

▲ 57

by a comprehensive test program on the 1/16 scale model of the tactical airplane.

It was the purpose of the report to describe the model and its characteristics and to show briefly how some of the flying characteristics of the full-scale airplane were simulated. A few simple measurements of the significant parameters were made for comparative calculations with the full-scale airplane.

Comprehensive coverage of the model testing was made on 16 mm motion picture film. Approximately six hundred feet of film was submitted with the proposal to the Navy. It was estimated that a showing time of one half hour was required with the projector operating at 16 frames per second. The test sequences in the film included the following scenes:

---

57) A 1/16 scale takeoff and landing platform was also constructed and subjected to testing. Both it and the L-200 model were tipped to extreme angles to demonstrate the static stability of the airplane in place on the platform.

58) A good view of the L-200 model and platform; the latter consisted of netted areas for takeoff and landing. The large lower net was designed to mesh with the pads on the tips of the empennage and the smaller net to one side was provided to catch the aft tip of the wing tip gun pod.

▲ 58

1. Operation of control handle and elevator.
2. Offset rudder and engine thrust and acute angle of flying wires to airplane centerline ensured tension in flying wires.
3. Thrust control achieved by use of two sets of ignition points (advanced and retarded) and an electrical relay actuated by operator through control handle and flying wires.
4. Scenes showing various methods of preventing models from crashing. A bamboo pole with two fishing lines to the wing tip pods was the first method used. The second method was to attach one fishing line to the outer wing lined up with the direction of flying wires.
5. With the bamboo pole, very small rubber bands were attached to fishing lines to prevent fouling in the propeller. Scenes showed that the tension in the rubber bands had a negligible effect on normal forces supporting the model.
6. Initial tests shown of hovering flight and controlled translation of the first conventional model prototype.
7. Extreme sensitivity due to the large elevator and its travel on the conventional model caused difficulty for the operator to make controlled hovering flights. After several modifications to reduce angular travel and the elevator area satisfactory control for hovering was achieved.
8. After preliminary testing of the conventional model a 1/16 scale model of the Lockheed L-200-1 was constructed.
9. Scale models of the landing, and takeoff platform proposed were made and static tipping tests were made to check in-place stability.
10. Some difficulty was encountered in oscillatory motions due to the bamboo pole method of safety and the coordination required between the pilot and the safety man.
11. The second method of safety devices consisting of an outboard fishing line with its line of action through the center of gravity of the model caused fewer difficulties with non-inherent airplane oscillations.
12. Scenes showing takeoff and landings from the floor and from the proposed platform.
13. Scenes showing the model hovering near the ground demonstrating good control effectiveness. Note that occasional contacts of the tail pods and the ground occurred.
14. Tests showing controlled translation back and forth in the flying model.
15. Scenes showing the model hovering in a side-wind velocity (normal to wing area) of approximately 6 to 7 kts. This corresponded to a 35 kt translation velocity of the full scale airplane. A blower was set up on the floor which created a jet of air approximately twice as large as the model planform at the distance from the blower corresponding to the 5 or 7 knot wind speed. Velocities of the jet were measured with a standard duct velometer.
16. Several transitions from vertical flight to horizontal were made after takeoff from the platform. Due to the small space available for testing and the limitations of the safety method, transitions from level to vertical flight were not attempted in this series of tests.

59) Stills from Lockheed's 16 mm film of the L-200 free flight scale model undergoing testing. It was determined from the tests that hovering control of the model was very good and that the full-scale airplane would be even better.

Description of the Model and Test Equipment

Conventional Model

This model was constructed from a stan-

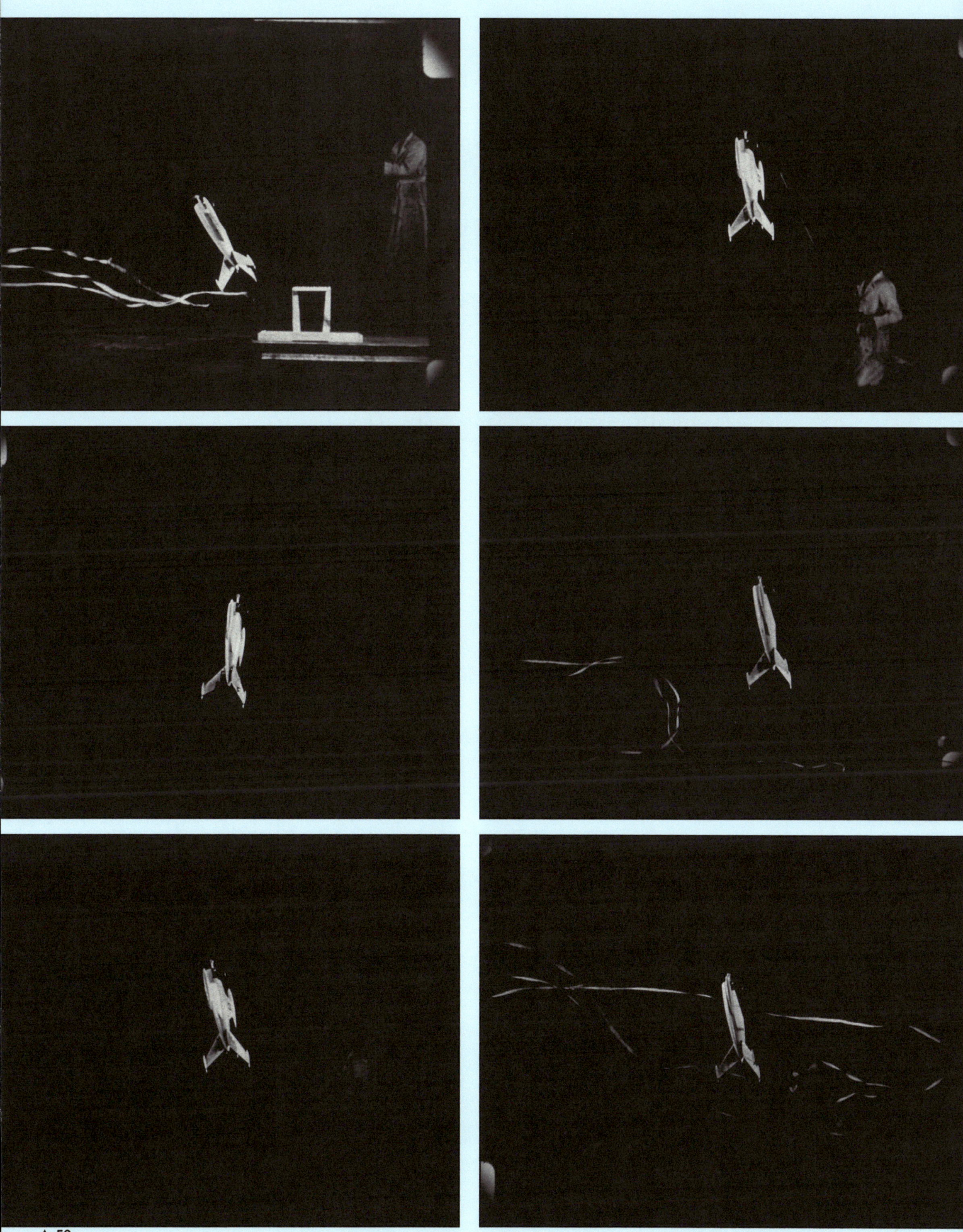

▲ 59

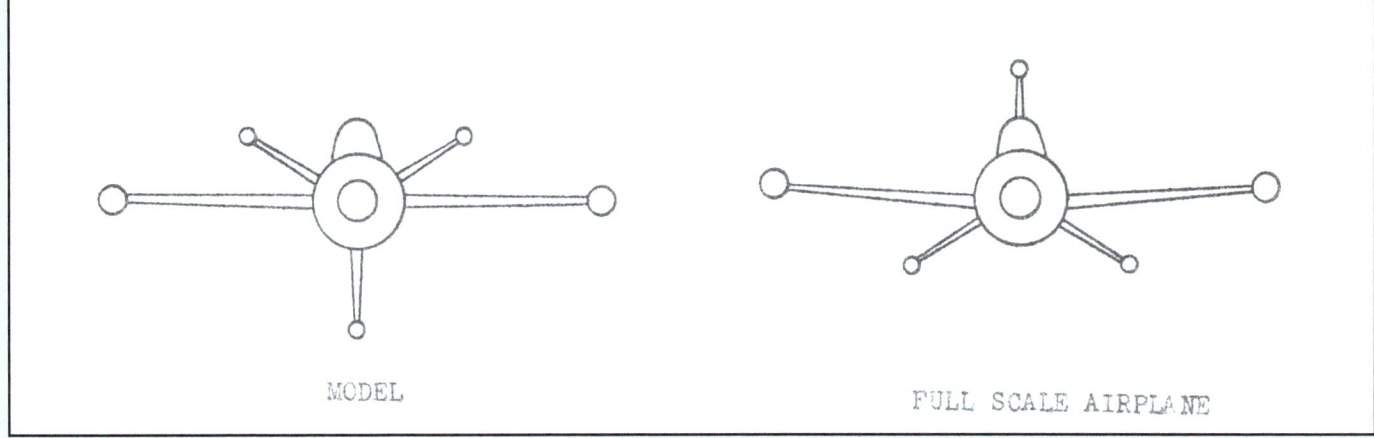

▲ 60

dard commercial kit and represented a conventional wing and tail arrangement. It was also equipped with a conventional landing gear although this was not used in this series of tests.

The weight of the model was approximately 34 ounces and the maximum thrust obtained was 38 ounces. Hence the thrust-weight ratio for hovering was slightly more than 1:1 in the majority of the tests.

A complete description of the model was not provided since it was not pertinent to the project except for check-out of the fuel system and power control.

L-200-1 Model

The 1/16 scale model was constructed in the early stages of design of the full-scale airplane, hence it did not represent exactly the final configuration. The wing configuration and the fuselage and body shapes were essentially correct and the empennage areas were accurately represented on the model. However, the empennage position was changed, moving the vertical surfaces on the full scale airplane from the bottom to the top, as shown in the drawing above.

This difference had little or no effect on the flying characteristics in the regimes which were tested since only pitch moments were of interest and the control surface effectiveness of the elevators was the same in either case in vertical flight and hovering.

General views of the model configuration are shown in the photographs on p. 67. These views show the side view and planforms of the complete model as constructed. It was found to be necessary in testing to remove the spinner and canopy section in order to attain sufficient engine cooling and thrust-weight ratios for hovering.

A 1/16 scale platform consisting of netted areas for takeoff and landing was also constructed for the model tests. A large lower net was designed to mesh with the pads on the tips of the empennage and a smaller net to one side was provided to catch the aft tip of the wing tip gun pod. Pictures of the model on the platform are shown on pp. 68-69 in which the platform was tipped to extreme angles to demonstrate the static stability of the airplane in place on the landing and takeoff platform. The design of the full-scale platform on the ship was actually gyro-stabilized such that the platform was level at all times. The reason that the airplane and model were so stable on the platform was that three widely separated axes of rotation existed between the tip contact point and the three empennage contact points. In order to rotate about any one of the three axes required side motion of the other two points which was effectively provided by the pre-load tension in the net cables. Lockheed noted that the model still held in place even after one empennage contact point came out clear of the lower net.

60) Drawing showing the configuration differences between the free flight 1/16 scale model on the left and the final proposal on the right.

61) Reconstruction of an earlier configuration of the L-200 based on the free flight model; compared to the plan on p. 14, the aircraft has an inverted tail and the wing has 0° dihedral.

62-63) Additional views of the original Lockheed L-200 display model. (John Aldaz Collection)

L-200 Early Configuration (Reconstruction)

Based on Lockheed 1/16 Scale Free Flight Model

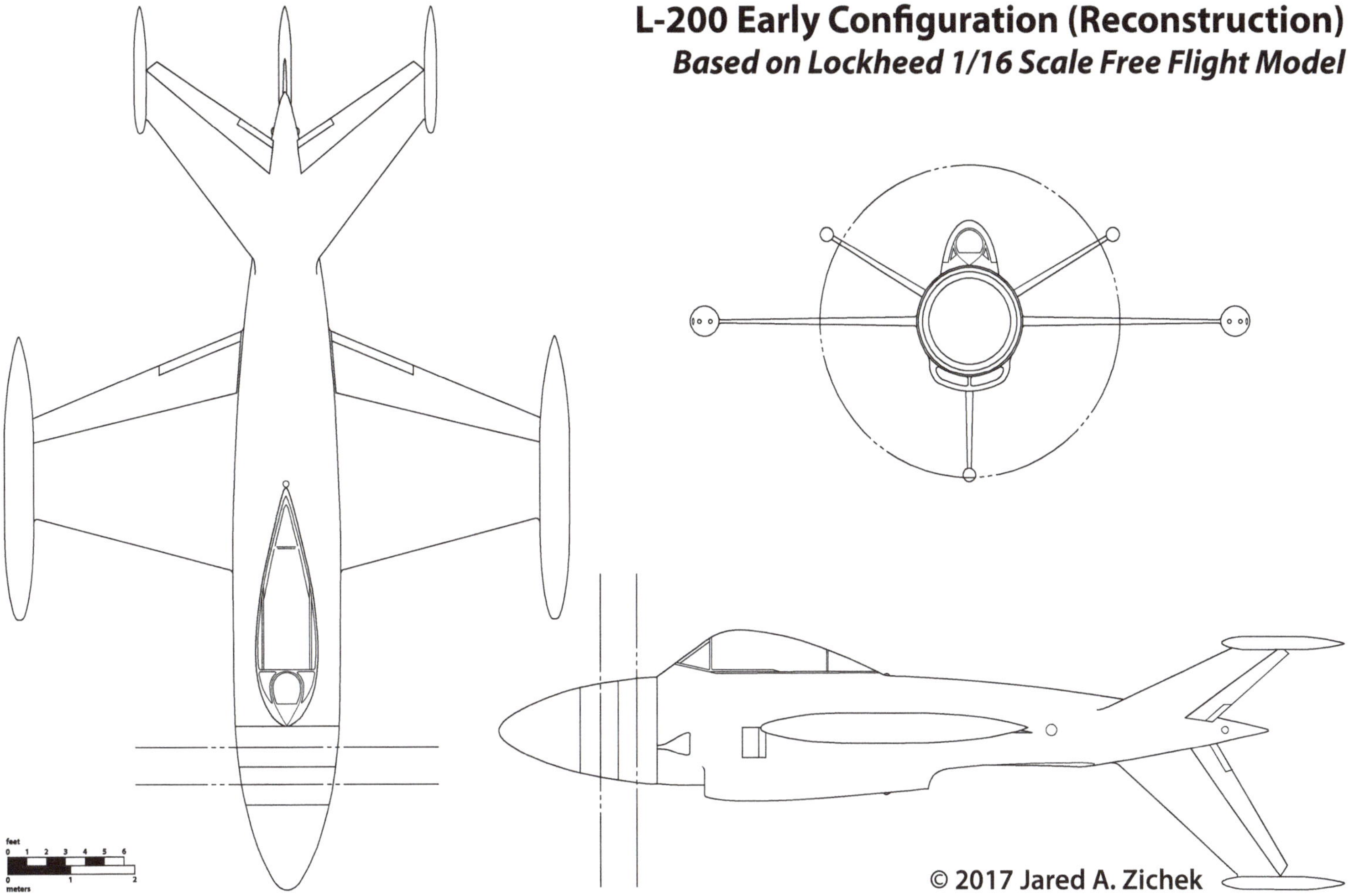

▲ 61

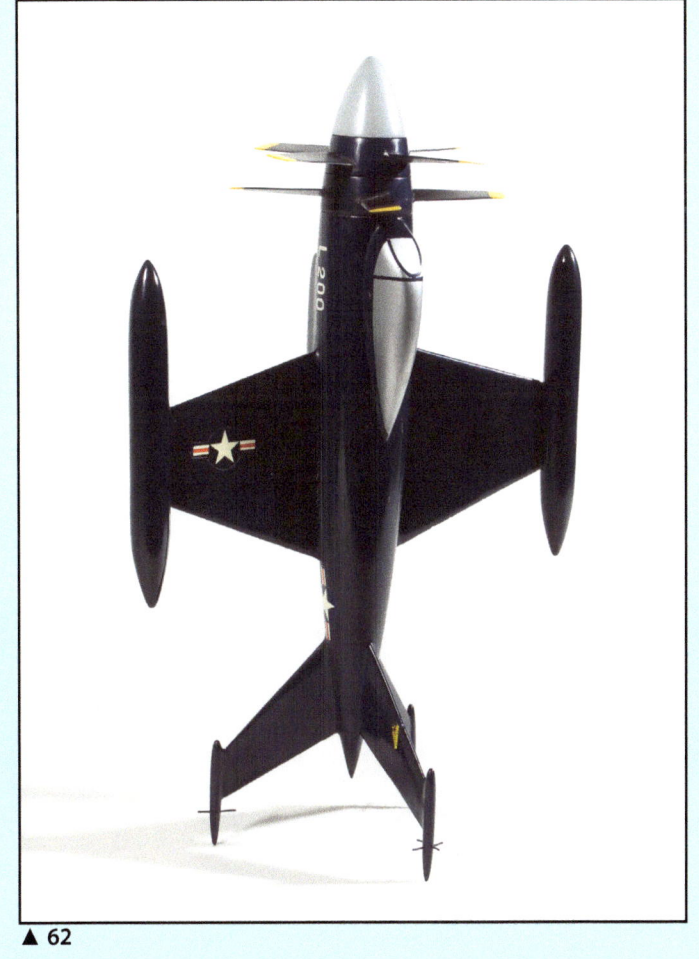

▲ 62

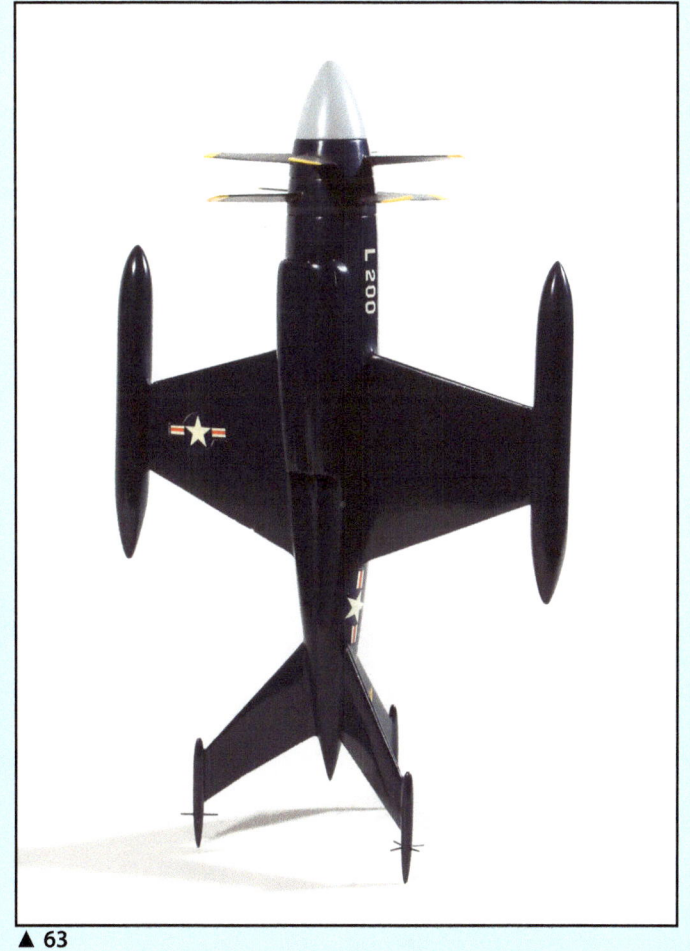

▲ 63

www.ingramcontent.com/pod-product-compliance
Lightning Source LLC
Chambersburg PA
CBHW041124300426
44113CB00002B/53